KU-263-578

Cardiff Libraries

www.cardiff.gov.uk/libraries

Llyfrgelloedd Caerdydd

www.caerdydd.gov.uk/llyfrgelloedd

CL

ACC. No: 02746550

STIRLING AND HOT AIR ENGINES

Designing and Building Experimental Model Stirling Engines

STIRLING AND HOT AIR ENGINES

Designing and Building Experimental Model Stirling Engines

Roy Darlington and Keith Strong

THE CROWOOD PRESS

621.42
DAR

First published in 2005 by
The Crowood Press Ltd
Ramsbury, Marlborough
Wiltshire SN8 2HR

www.crowood.com

This impression 2010

© Roy Darlington and Keith Strong 2005

All rights reserved. No part of this publication may be reproduced or transmitted in any form or by any means, electronic or mechanical, including photocopy, recording, or any information storage and retrieval system, without permission in writing from the publishers.

British Library Cataloguing-in-Publication Data
A catalogue record for this book is available from the British Library.

ISBN 978 1 86126 688 0

SAFETY WARNING

The reader accepts that the use of information provided in this book is done so at their own risk. The authors and the publisher do not accept and are not liable for any loss or damage which may be suffered from the use of this information. As in all engineering disciplines, dangerous materials, processes and techniques are used. Before commencing any engineering task, it is the reader's responsibility to read, understand and implement any safety recommendations and safety standards that may be applicable to the processes undertaken. The contents of this book are based on the accumulation of many years of experience and knowledge in diverse areas of engineering and represent the opinions of the authors. Every effort has been made to ensure the accuracy of the information contained in this book, but the reader, as is good engineering practice, should double check independently all information they use. The information should be regarded as a guide; to ensure complete satisfaction with the variability of the condition of engineering materials, the full standards for that material should be consulted.

Typeset by NBS Publications, Basingstoke, Hampshire

Printed and bound in Singapore by Craft Print PTE Ltd

CONTENTS

For the latest information and updates on this book and forthcoming titles please visit
www.stirlingengineer.com.

DEDICATION

I dedicate this book to my family who are always ready to listen, help and support me and especially to my wife Phyll who, although sadly no longer with us, gave me much encouragement throughout the years and always showed an interest in my engines even when they were on the kitchen table.

This dedication must be shared with Keith and Katie Strong. Without them, this book would never have been written.

FOREWORD

This book is not just about Roy Darlington's engines – it is about a personality who is closely associated with the widespread development of an engine at the level of a technical hobby for the home enthusiast, and while it describes how Roy developed this hobby it does not throw light on the person himself.

This book also brings two distinct and capable personalities together, each an expert in his own field, Roy and his co-author Keith Strong. Roy and Keith are a formidable partnership few can imitate.

For starters, the Stirling Engine Society and countless enthusiasts owe a debt of gratitude for the unflagging enthusiasm and intensity with which Roy has persistently promoted the Stirling engine. No single person has done as much to promote this interesting hobby.

Roy Darlington came on the scene in 1988 when he retired, having been active in model engineering since he was young and having built his first hot air engine in 1950. Not only has he not left the stage, he has extended his presence from the single annual exhibition in the early days to major exhibitions all over the country, plus countless other weekend exhibitions, college and school shows, and lectures both to the public and to various sectors of the industry at such places as universities, engineering museums and even London Zoo, with the occasional presence across the Channel, in France, Germany and Holland.

I first met Roy in 1989 during a return visit to the Model Engineering Exhibition in Olympia, after a gap of a few years. I remember very vividly the first time I saw him, patiently explaining the Stirling cycle principle to a number of onlookers. When they left I remained behind. Somehow in the friendly exchange he mentioned a book by James Rizzo. Quietly I said 'I am James Rizzo.' The look on Roy's face spoke a thousand words. We have remained friends ever since. Later Roy explained how he got involved in the hobby, after reading my book, *Modelling Stirling and Hot Air Engines* (Patrick Stephens Ltd, 1985).

If I am partly to blame for the zeal with which Roy took to the Stirling engine, I cannot help wondering what other kind of hobby would have caught his interest on his retirement – any other hobby would have been our loss.

For many decades the hot air engine experienced many ups and downs with the occasional revival, only to be overshadowed by concurrent development of some other source of energy – steam, electricity, the internal combustion engine. The same see-saw level of interest touched hobby enthusiasts and toy makers. Only the occasional article in a hobby magazine kept a fleeting interest. Then in the second half of the last century a number of articles began a revival which was, however, limited mostly to the readers. The real breakthrough came when Roy Darlington started exhibiting his models, as well as those of other contributors, first on the SMEE stand and then on the Stirling Engine Society's stand. However it was not just the engines that increased the onlookers' interest – it was Roy's constant presence and dedication that has spread the knowledge of the Stirling cycle principle a thousandfold.

But what about the man himself? Let us look at Roy Darlington as a person. Roy is an excellent craftsman – his models are a clear example of the highest technical level of ingenuity and perfection. Roy is a patient and courteous person. He goes to great lengths to explain how the Stirling cycle operates, and repeats the process as many times as necessary. Roy is one of the friendliest persons I have had the pleasure to meet and to know. His affability has endeared him to countless enthusiasts. Roy is indefatigable. At an age when most men take a back seat, he is always on the go, 'preaching the gospel of the Reverend Stirling', as he is well known for. His capacity to spend long hours on the various stands at shows is phenomenal.

While Roy is identified with his engines, the presentation and graphics and some of the technical writing were very much in the hands of Keith Strong. I have come to know Keith for some time now and few can appreciate as much as I do how hard Keith worked to make this book a 'special' one, and a very successful one too. The engines and the components would have stood up by themselves, but Keith's graphic and photographic interpretation have gone a long way to highlight them and to help readers not only to understand but virtually assimilate what Roy had intended in the first place. Keith Strong has the power to bring a humble component into the living room. He is the alchemist behind the scene – he does not steal the show, on the contrary, he enhances the capabilities of Roy Darlington. There could not have been a better twinning of two dynamic personalities.

It has been said that one of the best patrimonies a person can leave for posterity is a book. On that score I am sure that Roy's and Keith's book will also be a great gift to Stirling Engine enthusiasts for many years to come. There are rumours of some new books being written on this subject by Roy and Keith; I would anticipate that these will be eagerly awaited and if Keith's work in my soon-to-be-released new book *The Rev. Robert Stirling's models of the 'Air Engine'* is anything to rely on, I would expect a visual enhancement exceeding the standard of the first fine example.

James G Rizzo (Malta)
Founder member of the Stirling Engine Society
First Hon. President, The Stirling Engine Society
Author:
The Rev. Robert Stirling's models of the 'Air Engine' (Camden, 2007)
The Stirling Engine Manual: Volume 1 (Camden, 1995)
The Stirling Engine Manual: Volume 2 (Camden, 1999)
Modelling Stirling and Hot Air Engines (Patrick Stephens, 1985)

Introduction

ROY DARLINGTON

Born in 1924 in Worthing, West Sussex, Roy Darlington started constructing things at an early age. At the age of five he made his first electric light circuit powered by a battery. Throughout his childhood he made numerous model aircraft, fireworks and water turbines that ran generators to power his electrical experiments. He also belonged to the Worthing Model Aircraft club.

Whilst still at school, Roy joined the Air Defence Cadet Corps, which later became the Air Training Corps. There he learnt drill, aircraft spotting, started his education on aircraft engines and participated in training exercises on the South Downs.

When he left school in 1939 he was immediately mobilized and his first job was to receive child evacuees arriving at Worthing central station. His uncle ran 'Darlington Mushrooms' from some buildings at the back of the station and these were used to sort and issue rations to the children before they were placed with a family in the town.

Once his basic RAF training was completed, Roy was posted to Cosford, a technical training college. Roy wanted to be a flight engineer, but this was not possible due to being colour-blind so he trained as an engine fitter. Bomber Command had a very high casualty rate so it is quite likely that being colour-blind saved his life. His training involved basic engine fitting, drilling, riveting, then stripping and rebuilding engines. He worked on Gipsy Minor, Gipsy

LAC Roy Darlington during his RAF service.

Major, Kestrel, Merlin and radial engines. Then he progressed to the airfield and worked on Blenheims, Wellingtons, Hurricanes, Spitfires, Fairey Battles and twin-winged aircraft. At the end of the training he was given written and practical tests and then passed out, becoming an AC2 engine mechanic.

Roy's first posting was to the 611 fighter squadron at Biggin Hill in Kent, where he worked on Mk IX Spitfires. This was the cream of the RAF. The group captain of the squadron was 'Sailor' Malan, the famous air ace. It was a dawn to dusk job, with sweeps being sent to France twice a day or more. The squadron on the other side of the aerodrome was the Free French Cross of Lorraine squadron. The two squadrons would fly together, but whereas the 611 Squadron flew in three rows of four, the French squadron would fly in the Cross of Lorraine, so that when it flew over France it would be recognized as free French.

In 1943 Roy was posted to 149 Squadron, a bomber squadron at Lakenheath in Suffolk. Here the runways were the longest in the country and there were Nissen huts spread over a four-mile area. The base was known as 'little Libya', because it was very sandy, extremely cold in the winter and very hot in the summer.

The squadron consisted of Stirlings, a rather daunting prospect as the engine of a Stirling was 20ft above the ground. The squadron was laying mines in the fjords of Norway to attack the German navy, bombing Italy and bombing the rocket sites on the French coast. It was easy for Roy to guess where the aircraft were going by the amount of fuel and bombs they had to load. If they were flying to Italy then he could expect the next day off as the aircraft stopped in North Africa to refuel. Sometimes he would be transferred to another air base and one of the Stirlings would be chosen to drop spies, new recruits and equipment for the resistance, such as grenades and sten guns.

When a new squadron was formed at Chedburgh in Suffolk, Roy was amongst a group who were taken to become fitters and riggers. This was the 1653 Heavy Conversion Unit that trained aircrew who had been flying twin-engined aircraft to fly four-engined bombers. The crew would consist of a pilot, navigator, flight engineer, bomb aimer and gunners, who would fly together for five or six weeks doing circuits and bumps and bombing on a bombing range in Scotland until they passed out.

There were twenty-four Stirling bombers flying day and night. Roy's longest time in the RAF was spent here. It was hard work, forty-eight hours at a time, and he had to inspect and refuel the aircraft that were flying round the clock. At times, they were attacked by German bombers at both Chedburgh and at Lakenheath.

Eventually the squadron was transferred to North Luffenham in Rutland. It was a peacetime base with proper buildings and barrack blocks. It was here that Roy met his wife, and later his son was born there. Just before the war ended, more manpower was needed for fighting in Germany and across Belgium and Holland and a lot of airmen were transferred to the army as soldiers. Roy was probably too useful in his job to become part of this, so was luckily never put on the list. He loved the job and really enjoyed finding out what was wrong with an engine.

As the war continued with Japan, many RAF personnel were transferred to the Fleet Air Arm. Before reporting to Blackpool to join the Navy, Roy had to go to the RAF medical unit to see the medical officer. Looking at the records, he noticed that Roy was colour-blind and explained that he would be expected to work as a mechanic below deck with artificial light and would not be able to see the colours on the pipes. At the time Japanese aircraft were kamikazeing the carriers and the warships, so once again his life may have been saved by being colour-blind.

In 1946 Roy was posted to the 82 Photo Reconnaissance Squadron in West Africa, whose job it was to map the Gold Coast. After a boat journey to Takoradi in West Africa, followed by a twenty-four-hour train journey through the jungle, Roy arrived in Accra, where he was set to work servicing Lancasters. He must have done his job well, as Lancaster no. PA474, one of six Lancasters on 'B' flight that Roy maintained, is still flying to this day in the Battle of Britain Memorial Flight.

PA474 is one of only two Lancasters remaining in airworthy condition out of the

Relaxing in West Africa at the end of World War II.

7,377 that were built. She was built in Chester in mid 1945 and was earmarked for the 'Tiger Force' in the Far East. However, the war with Japan ended before she could take part in any hostilities so she was therefore reassigned to photographic reconnaissance duties with 82 Squadron in East and South Africa. While operating with 82 Squadron, PA474 had her turrets removed and carried the identification letter 'M'.

After Roy's service in West Africa ended he was sent home on a boat for demob.

During his four and a half years in the RAF he served in Training Command, Fighter Command, Bomber Command and Coastal Command.

In 1947 Roy moved back to Worthing and started working at Allen West in Brighton, where he continued to work for 42 years. He started as a fitter on the bench making switch gear and continued this for several years. He then started doing experimental work for the Admiralty, working alone in a Nissen hut. The project was semi-secret and involved trying to make switch gear shockproof so that it wouldn't trip out when the Navy fired guns. He honed and lapped Admiralty switch gear cylinders (dash pots) to give a delay of two and a half seconds to stop the shock knocking the contactors out. This was early training for making good-fitting pistons and cylinders.

At the same time, he was doing model engineering, making small steam, diesel and gas turbines. He became an outside fitter working in chemical works, factories, railway sidings, harbours, fuel installations and power stations all over the country. Eventually he was asked to form his own department and all the push buttons made at Allen West were collected together in that department. The department grew rapidly. He became a charge hand and then a foreman with about 300 people working for him. When an American company took over he was put in charge of internal and external transport.

Before taking early retirement in 1988 he spent a year as production controller working with computer information all day long – which he hated. He was much happier on the factory floor talking about problems and keeping flow lines going.

On his retirement Roy abandoned all his other hobbies and concentrated on hot air engines. He started gathering useful pieces of material from many sources. He joined the Worthing Model Engineers and found that he was the only one in the club making hot air engines. Everybody else was making steam engines or traction engines and there was a great deal of interest in the engines he made. He made engines to his own design from the outset and started doing exhibitions, getting crowds around him all the time. He started giving talks to Scouts and other organizations and found that he could interest people in the subject. He was asked to take his engines as part of a display and to give a talk at a medical show at the Assembly Hall in Worthing. It was a great success and he started doing medical shows as well as shows for the Model Engineers.

In my workshop. (Reproduced with the kind permission of Mike Chrisp)

ABOUT THE CO-AUTHOR

Keith Strong.

The shows he has participated in include The Dome at Brighton, The Brighton Centre, RAF Hendon, London Zoo, Beaulieu, Harrogate, Croydon and Alexandra Palace, which was later moved to Olympia and then Sandown Park. Every year there were shows at Picketts Lock and Wembley. He has also done shows in Bristol, Bath, Southampton, Southampton Brickworks and all sorts of clubs over the south coast. He has participated in shows at many universities including London, Surrey and Brunel. At the Society of Model Engineers centenary show at the Brunel University it was Roy's job to show the Duke of Gloucester around. Every year, for the last eleven years he has exhibited his engines at Flers and other places around Normandy.

Roy helped to form the Stirling Engine Society in 1997. He has had articles in magazines all over the world including France, Australia, Germany and Canada. He corresponds with people worldwide and has become an authority on hot air engines. His enthusiasm on the subject is infectious.

Keith Strong was born in 1965 and grew up in Chichester, West Sussex. His interest in mechanics and engineering began when he and his father made scale models together. This led to the start of a lifelong fascination with aircraft. His first engineering accolade was to achieve one of the highest ever scores in his school metalwork examination for designing and constructing a bicycle frame to 'Tour de France' specification using lightweight alloy steels. At the age of eighteen he studied landscape construction at Merrist Wood Agricultural College in Surrey where he broadened his engineering talents working on agricultural and horticultural machinery. It was then that he also developed his continuing passion for the restoration of old cars, especially Minis and Triumphs.

He has trained in and studied Motor Vehicle Technology, Engineering Design and Draughting, Computer-Aided Design and Architectural Engineering Construction,

AutoCAD, Pro Engineer, Inventor, Solid Edge and microprocessor programming and control at places such as Chichester College of Technology, Queen Elizabeth's Training College in Surrey, Hatfield Polytechnic near London and the University of Brighton. He undertook further training at Jaguar Cars Ltd where he learnt hands-on about robotics, jig tool and gauging fixture design on V12 engines and 'Body in White' design on the XJS and XJ40.

Keith's broad spectrum of employment has included: the installation of computer-aided design systems, designing television and broadcast equipment, writing and designing websites and network installations. He also worked as a CAD designer for the Surrey Space Centre at the University of Surrey. Here his projects included UoSat-12 (a research satellite launched from Baikonur in Kazakhstan), ground support equipment, minor jet propulsion motors, total loss propulsion systems, instrumentation such as optical cameras, earth horizon and star cameras, magnetometers, magnetic torquers and solar panels, and other NASA research satellites.

Keith has undergone several episodes of pioneering surgery on his knee where the bone has been replaced with titanium, but these setbacks have never dampened his enthusiasm and he often jokes that he feels that he has engineering in his bones.

His interest in cars led to involvement with a local motor club where he since has won the autotest championship. He excelled in rally navigation and is a trained rally marshal in fire fighting and rescue with experience as Clerk of the Course, Sector and Chief Marshal on both international, national and local events. During this time he built and campaigned a Group 'N' rally car.

Keith is a natural inventor and constantly thinks of new ideas and improvements for existing designs. Keith and Roy met at a model engineering show and his fascination with the subject of Stirling engines grew rapidly. He has since been researching and designing multi-cylinder Stirling engines and occasionally assists with shows and presentations. Both attended the International Stirling Engine Conference in Osnabrück in Germany. His recent projects include the building of a ground effect hovercraft. He married Katie in 1995 and their son Curtis was born in 1999.

Keith has given unstinting help in writing this book. After being approached by The Crowood Press, as the ideas and content began to formulate in my mind, I was confident that Keith was the one person I knew who possessed the right mix of talents, technical knowledge and skill who could aid me in the co-writing of this book. Without hesitation, he agreed and over eighteen painstaking months, he has ably interpreted my words into script, drawing from his own knowledge of the subject, and his knowledge in other engineering disciplines. His strengths of patience, exceptional attention to detail, excellent CAD and computing skills and his dedication have transformed my work of many years into print. He generated the engineering drawings, 3D computer-generated models and sketches on the computer from actual parts and took the majority of the photographs that appear in this book. Over the months spent together, a great partnership developed of like minds and an enduring friendship will continue long after.

Roy Darlington

1 A Brief History of the Stirling Engine

HEAT ENGINES

A heat engine is a term that can be applied to all engines that produce any sort of work from energy. They have a direct or indirect by-product of heat; even some entities that may not appear at first glance to be any sort of engine can involve the heat to energy process and thus be loosely classified as a kind of heat engine.

The book entitled *The Evolution of the Heat Engine* by Ivo Kolin gives a fascinating insight into many examples of heat engines and, for those who wish to look into this area further, I can think of no better starting place. The variations that come within the definition of a heat engine initially may be surprising or a stretch of the imagination, but all work on the principle of producing energy from heat or vice versa, or having a by-product of heat as a result of converting energy. Here are some thought-provoking examples:

- animals
- batteries
- cannons
- Carnot engine (cycle)
- cold air engines
- diesel
- gas turbine
- guns
- hot air balloon
- jet
- Newcomen atmospheric engine
- NSU/Wankel
- nuclear power
- rotary engines
- Papin vacuum engine
- photovoltaic cells
- refrigeration
- rockets
- Savery engine
- steam turbine
- Stirling.

HOT AIR ENGINES

Hot air engines are indeed a type of heat engine and are engines that use air as their working medium. They work by drawing a fresh charge of air on each cycle and exhausting that air during the cycle (open cycle). One of the drawbacks of this type of engine is that much of the heat used in the cycle is typically discarded in the exhaust.

STIRLING ENGINES

Stirling engines come within the definition of hot air engines, but differ as a select group by having the working medium (typically air or helium) enclosed within the engine and use that medium in a reversible closed heat and pressure cycle.

MILESTONES RELATING TO THE DEVELOPMENT OF STIRLING ENGINES

Experiments on hot air engines were documented prior to Robert Stirling's patent of

13

1816. George Cayley (1773–1857) of Yorkshire and sometimes called 'Father of Aviation' was a notable innovator who published details of his own experimental hot air engine in 1807. He referred to current experiments in France and commented that the idea was by no means new in England. His hot air engine was different to a Stirling engine in that it drew air through flames and hot coals rather than through a regenerator and it used an open cycle, drawing in a fresh charge of air on each cycle.

Sir George Cayley's Engine, 1807.

The Reverend Robert Stirling. (Reproduced with the kind permission of East Ayrshire Arts and Museums)

The air was expanded by forcing it through the flames, causing the piston to move downwards, which in turn drew in a fresh charge of air. The idea was never successful, which George Cayley attributed to a copper piston that was made of irregular copper sheet and therefore not air-tight. The idea was not patented until 1837, when he suggested it might be suited for use in a road carriage.

Robert Stirling was born in Perthshire on 25 October 1790, one of eight children. It is likely that an engineering interest came from his grandfather Michael Stirling, who had been responsible for the invention of the first rotary threshing machine in 1756.

James Stirling, Robert's brother, was a good engineer and played a major role in the Stirling engine patents and development and was more than able to demonstrate and explain the working of their closed cycle hot air engines to such eminent people of the time as Sir George Cayley and Sir Robert Stephenson.

At the age of fifteen Robert attended Edinburgh University where he studied Latin, Mathematics, Greek, Logic and Law. He went on to become a Student of Divinity at Glasgow and Edinburgh Universities and was ordained as a minister of the Church of Scotland in 1816.

Robert Stirling's first patented hot air engine included the regenerator, which he called an economizer. This is an important part of modern Stirling engines, in which ultimate heat efficiency is required. Robert Stirling was reported to be concerned about the death toll

Robert Stirling's engine of 1816.

of steam engine operators and bystanders, where the boilers on the steam engines sometimes exploded, and he realized that hot air engines could take their place, undertaking the tasks more safely as hot air engines at the time didn't operate under the same pressures. His patent applied for in 1816, entitled 'Improvements for diminishing the consumption of fuel and in particular an engine capable of being applied to the moving machinery on a principle entirely new', deals mainly with the principles of the regenerator. These engines did not actually take his name until the 1900s. In 1818 one of Robert Stirling's hot air engines was used as a quarry pump and ran for two years, although, as with many of the early engines, it suffered problems due to the poor quality of metal and materials available at the time, which were unable to withstand the high temperatures.

In 1819 Robert Stirling married Jean Rankin and they had seven children, four of whom also became engineers. In Kilmarnock he worked together with Thomas Morton who provided workshop facilities. They were both interested in astronomy and invented several optical instruments. In the 1820s through to 1840 he also worked closely on experiments and improvements with his brother James and they obtained further patents in 1827 and 1840 for improvements to hot air engines.

Rev. Stirling died at the age of 88, by which time he had become 'Father' of the Church of Scotland. Two of his engines survive, one housed at the Royal Scottish Museum and one at the University of Glasgow.

In 1824 M. Sadi Carnot published a book with details of thermodynamic principles that became known as the 'Carnot cycle', which introduced the idea that the efficiency of these engines relies only on the temperature difference and not on the fuel that powers them. The theory remains the most effective way to convert heat into work.

In the nineteenth century, many Stirling engines were produced and sold commercially as alternatives to steam engines. In particular, John Ericsson (1803–89) produced and sold thousands. He was born in Sweden and moved to England in 1826, bringing with him a working model that he called a 'flame engine'. This was an open cycle engine that was powered by wood, but he found it to be unsuccessful using coal because the heat was too intense, so the experiment was abandoned. In 1833 he developed a closed cycle engine with external heating which he called a 'caloric engine', a name by which many Stirling engines were known in the nineteenth century, also sometimes referred to as hot air or Carnot cycle engines. It generated 5bhp, had two double-acting cylinders and a tubular regenerator. Between 1838 and 1850 he designed many hot air engines with wire gauze regenerators. In 1851 his 'Caloric Ship' was built which was a 260ft ship powered by a large version of his caloric engine. It had four cylinders each with a stroke of 6ft. During trials the ship never achieved the speeds predicted, making it unsuitable for commercial purposes. Unfortunately it sank in a storm whilst still undergoing tests and improvements, and after raising was fitted with a steam engine. In 1873 he developed a closed cycle solar engine which was patented in 1880.

In 1937 NV Philips, a Dutch electrical company, was looking for an efficient engine to power radio receivers and started an extensive Stirling engine development programme that lasted for many years and included automotive, solar, cryocooler and submarine applications. Philips produced quiet, efficient engines, some of which were capable of up to 5,000rpm and 38 per cent efficiency. In 1958, General Motors

took over the company and eventually the programme was terminated.

In 1964 William T. Beale produced a free-piston engine with only two primary working parts. This type of engine is often called the 'Beale' engine. With a team of people he founded a company called SUNPOWER in Athens, Ohio, in 1974. The company does excellent work in applying modern technology, materials and a sensible design ethos to the free-piston Stirling engine.

In recent years. Stirling engines have been studied at NASA for use in powering space colonies with solar energy. They have also been used for air circulation and refrigeration. NASA has also designed and built engines that have been fitted in all types of vehicles, halving their fuel consumption. Infra-red sensors in guided missiles use Stirling cryocooler technology to keep the sensors cool. This makes the infra-red sensors sensitive enough to be useful.

Certainly all of the major motor manufacturers have considered this technology and some have built or funded various vehicles that use Stirling engines. It remains to be seen if enough research and development effort will be employed to adapt the Stirling engine to become a suitable and viable alternative.

Predictions are always dangerous, especially in books, but certainly Combined Heat and Power (CHP) will make a real impact in the way we use and generate electricity.

The Stirling engine's ability to use almost any type of fuel, particularly when considering hydrogen and solar, means that it is almost assured a major role in an oil-depleted society.

2 Principles and Performance Enhancing Techniques

It would be remiss of me not to start by recommending two books on this subject. They are *Engineer to Win* by Caroll Smith and *The New Science of Strong Materials, or Why You Don't Fall Through the Floor* by J. E. Gordon (details in the bibliography). They are written in a language you can understand and fit well with my own ethos. You would do well to enhance your engineering knowledge by tracking them down and reading them. There will be the inevitable similarity in some areas partly due to the subject matter and partly due to the influence the books have had on my engineering. One further book that is good for brushing up on your basic workshop techniques is *MetalCraft Theory and Practice* by John R. Bedford.

In a Stirling engine there are two main components that are key to its successful operation. It is necessary to know what these components do in order to understand how they interrelate to provide some sort of prime movement or power. This section will help you to understand what a Stirling engine is all about. There is also a wealth of information in this book that will help towards creating smoother, more efficient and better engineered engines. Chapter 16 concentrates on this in particular.

The two primary parts of a Stirling engine are the power piston and the displacer piston.

The power piston provides the motive power typically to turn a flywheel and the displacer piston is used in the generation of the forces to provide that motive power. For this reason the displacer piston will be discussed first.

THE DISPLACER

The displacer assembly primarily and usually consists of a piston and cylinder. The cylinder is normally heated at one end and cooled at the other. In order to simplify this slightly to understand how it works, it would be useful initially to consider two processes.

Process One

If the air inside a cylinder is heated the pressure inside will increase.

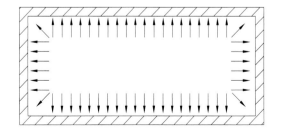

A pressurized cylinder (heated).

Process Two

If the air inside a cylinder is cooled the pressure inside will decrease.

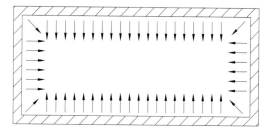

A depressurized cylinder (cooled).

A displacer piston's function is to move or displace the air from one end of the displacer cylinder to the other. This means that the displacer piston must allow the air to pass by and is usually engineered as a loose-fitting piston for this reason.

Hot and cold ends of the displacer cylinder and the power piston reaction.

When one end of the displacer cylinder is heated and the other end is cooled there are two temperature zones, one available at either end of the same cylinder. The displacer piston as it is moved from one end of the cylinder to the other transfers or displaces the air to the opposite end of the cylinder. That air will now be heated or cooled depending on which end of the cylinder it has been displaced to, where the appropriate pressure reaction takes place.

THE POWER PISTON

The pressure changes that are produced in the displacer cylinder are also transferred to the power cylinder and act on the power piston, causing it to move in and out, or reciprocate, within its cylinder.

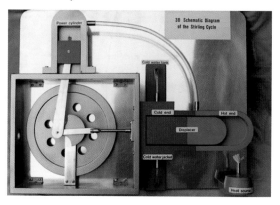

A model that was built for demonstration purposes gives a good indication of just how simple a Stirling engine can be. (Reproduced with the kind permission of Graham Darlington)

The simplest drive to these two pistons is often taken from a flywheel. The connections to the displacer and the power pistons are typically made 90 degrees apart. It is of no consequence which way the 90-degree phase is, as it only determines in which direction the engine turns the flywheel. The 90-degree phase must, however, be led by the displacer cylinder.

Testing and observation have shown that the shorter the transfer from the displacer cylinder to the power piston, the better the engine performs. This volume of gas contributes to the dead space. This is highlighted with three types of Stirling engines.

There is the Beta type, in which the power piston and the displacer work back to back in the same cylinder.

The other two types are the low temperature differential Ringbom and the Low Δ (Delta) T, in which optimum performance is required in order to operate at extremely small temperature

Beta piston assembly showing power piston and displacer piston as they would be inside the cylinders.

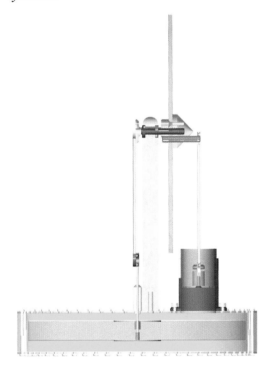

Cutaway view of a Low Δ T piston assembly.

differences. In these engines, the two cylinders are also back to back.

Other Stirling engines use a simple transfer tube such as a slot or hole in the casing.

Separate mounting of the two cylinders is possible.

METHODS FOR INCREASING PERFORMANCE

The pages that follow give a detailed description of all the methods that I use to improve the performance of the Stirling engines that I build. There are a few instances where I have covered areas that I do not use because they are usually unnecessarily complex for model Stirling engines. They have been included to reveal some of the further potential improvements that can be made.

Regeneration

Regeneration is a method used to make the working gas (in most cases air) give up some heat as it moves from the hot end of the displacer cylinder to the cold end and then to regain that heat as it moves from the cold end back to the hot end. What happens here is that the change of temperature in order for the Stirling cycle to take place is reduced. In other words, the engine will be more efficient because it will need less cooling or heating for the same power output. In reality, the actual temperature difference is increased because the regeneration process enhances the temperature variance.

The trade-off when using a regenerator is that you are deliberately creating an area of dead space, which will have some detrimental effect on efficiency, but this will be compensated for by the overall improvement in the engine's output.

Regenerators can be made from a variety of materials. For model engines wire wool or gauze are often used, as are serrated or perforated metal sheets. More complex methods utilize a metal or ceramic matrix. The regenerator material is placed in the air transfer section of the displacer. In the case of perforated or serrated metal it can even be wrapped around the displacer piston, or it can be set into a channel cut into the displacer cylinder. The wire gauze, metal or ceramic

Cutaway view of a displacer cylinder with regenerator material.

matrix can be set into a space between the outside of the displacer cylinder and a secondary tube which forms a twin-walled cylinder. Usually I would build an engine without a regenerator where the gas simply uses the gap between the piston and cylinder to get from end to end.

When thinking about the design of a regenerator it would be useful to build-in a degree of turbulence, as this increases the amount of heat transferred to and from the regenerator material. This turbulence happens to a certain extent by default.

For the engines I have built, I have not utilized the potential gains a regenerator might return but have concentrated more on good design, accurate workmanship, reduction in weight and reduction of frictional losses. I find that this approach gives a simple engine that

performs well without the added complexities of building a regenerator into it.

LOSE WEIGHT AND REDUCE STRESS

Some of my engines run at up to 3,000rpm and on occasions the power generated from these relatively small models has bent $^{1}/_{8}$in aluminium plate. It can be seen that some serious forces are at work, making designers and engineers to use trade-offs between the need to contain those forces while keeping weight to a minimum. The high-speed rotation and reciprocation require the materials and design of components to be as small and light as possible, but to counteract the effects of bending frames and other supports we would aim to make the construction heavy enough to withstand the loads. As with most engineering solutions, there is always a compromise to be had. In general, the trade-off will be just enough strength (usually with a safety factor applied) versus making components as light as possible without failure.

Many of us will not need to do detailed stress analysis or worry too much about fatigue strength. While analysis of strength of materials undeniably has its place, some model engineers work with materials they have been given or have found. Reliable data on the type and condition of the material will often be sketchy and to do calculations based on that material will be starting at best from an educated guess. Even identical materials that have been subjected to different heat treatments can have markedly different properties. However, there are some simple techniques that are worth consideration in helping to maximize the strength of any component while minimizing its weight.

Holes, radii, steps, notches, tool marks, flanges, joints, poor machining and corrosion are all features that reduce the life of components. Many of these are essential features of a part's design, but with a little thought and some simple machining it is

Stress concentrations and flow for various machined features on a shaft.

possible to recognize and minimize their detrimental effects.

Holes

When designing a part that requires holes, it is preferable to leave at least one and preferably two times the diameter of the hole between the edge of the hole and the edge of the component. When drilling holes always use sharp drills and if possible drill at slow speeds; if drilling at high speeds use cooling solutions. This will reduce the risk of altering the crystalline structure of the metal through localized heating. This heat often makes metal brittle just at the point where it needs to be malleable. On components that are bolted together, ensure that the holes line up properly; do not bend and stretch the metal to make it fit. Holes should be deburred, if not chamfered. If shafts or pivots are to be used make sure that an appropriate bearing or bush is used. When choosing the size of a hole it is wise to refer to the relevant standard for tapping and clearance, because a hole that is too large will eventually loosen, fret and cause other damage. One method of peening (hardening) the hole edges is to place a suitably sized ball bearing in the chamfer and hit it with a hammer. This peening compresses the surface of the hole and helps to stop cracks from propagating. Instead of drilling holes to lighten a component, consider reducing the cross-sectional area. However, despite all of the above methods the hole will still be a weak point. It is better to remove a weak point if at all possible and design the structure in a different way.

Radii

When thinking about the size of a radius in a component use the largest reasonable size, because the bigger the radius, the less the stress and the longer the component will last. If an internal corner of a blind slot is square, consider using a radius instead as this reduces the tendency of the metal to crack. The sharper the corner the more likely the component is to crack. In the case of a simple clevis it is as easy to radius the slot as it is to leave a square end to the slot. Always blend a radius into its surrounding edge, as not doing so will introduce a weak spot.

Steps

If you can, avoid using steps. Obviously there are times when this will not be possible. If you are machining a step on to a shaft and the step has to be square, consider undercutting the internal corner with a small radius. This will produce a stronger part with just a simple machining operation. It is possible to use a radius instead of a step by using a collar with an internal radius that fits over the shaft. The radii of the shaft and the collar can be placed back to back, which gives a stable square edge on which to mount a bearing. This also maintains a radius to step transition on the shaft.

Notches and Tool Marks

Notches can occur through a variety of causes, from corrosion to tool marks. Any component that is subject to repeated stress may suffer from 'scratch or notch propagation'. This is where the repeated bending and flexing of a component seeks out a weak point in its structure and it starts to crack from there. Always finish components with a good surface finish – deburring, radiusing and chamfering may take more time, but will increase the longevity, strength and appearance of most components.

Flanges

When a part is flanged to a tube it is a good idea to put a radius on the non-contact side. It is also worth considering machining a counterbore in the part it is attached to, in order to locate the flange accurately.

A simple flanged cylinder.

Joints

The fewer joints used, the more homogeneous an item will be. Most joints will need to be rigid and it may be possible to machine a slot and let in one side of the joint to improve its location and security in addition to bolting it together. In a highly stressed part that has been joined together by heat there may be a requirement to anneal it so that the metal is not brittle. A part that is subject to violent changes in direction may also need to be annealed to cope with the repeated deflection it may suffer, though it might be better to revise the material

choice in this instance for a material that does not work-harden so easily.

Any joint should provide suitable location and be neither too big or too small. Too big and the assembly will fit loosely, putting undue stress on a component when bolted together. Too tight and the part may be damaged, and may also damage the parts to which it is fitted. If possible when considering the design of parts that are bolted together, try to thicken the cross section at the point where the bolt holds the parts together. Try to avoid the use of woodruff keys as this weakens any shaft; it is much better to use a non-invasive technique such as a collet arrangement (*see* Stirling Silver II, Chapter 12).

Methods to create a thickened section where bolted joints may be required.

Poor Machining

Deep scribing marks, poor handling, rough turning or milling, excessive grinding or hammering and the use of blunt cutting tools all reduce the life of any component. However, in the real world we all drop components or scribe a line in the wrong place, so try to allow sufficient material to finish a component properly. Linish if necessary and try to obtain the best quality finish possible. The extra care taken here will prolong the life of your components and make the job look more professional. If you wish to make a part as

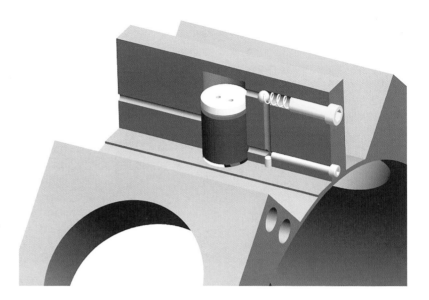

This cutaway view shows the air intake, piston and seal, compression chamber, one-way valve and pressurized air transfer tube that form a method of pressurizing a crankcase.

strong as possible, it is good practice to achieve a high surface finish both internally and externally. This is the practice in top-level engineering such as Formula 1, so should be worth some consideration for your own parts and can increase the life of a component by up to 30 per cent.

Corrosion

Avoid using metal that has corroded, as the corrosion will have chemically altered some of the structure of the metal and may penetrate areas that cannot be seen. When a Lancaster I worked on during World War II was being serviced, I found that a white powdery deposit had appeared through the black paint on the underside of the wing. On examination it was found to be inter-crystalline corrosion which had to be cut out and patched. This is a prime example of the kind of corrosion that can be introduced during manufacture or storage.

If using tubular components, use a corrosion inhibiter or oil on the inside surface. Take care not to induce galvanitic corrosion between dissimilar metals. Plating of metals is helpful and it is possible successfully to electroplate a variety of metals using simple home-based

techniques. Anodizing, for example, gives a hard glass-like surface that also protects against corrosion. Painting of components is also an easy and cheap way of inhibiting corrosion.

PRESSURIZATION

Most commercial Stirling engines will use some form of pressurization. One of the most prolific producers of Stirling engines was the Dutch company, Philips, whose engines would use pressures from 15 to 220bar to improve performance. Pressurization of Stirling engines may be unnecessary for general model engineering, but if it is performance you are after it is a method well worth considering. Pressurized Stirling engines will give a power equivalent similar to their 'Otto-cycle' counterparts.

It is likely that a method of automatically maintaining the pressure levels will be needed. If using a constantly recharging method of pressurization, then a suitable pressure regulation valve must be used. A prime Philips method was to use a small piston which compressed a tiny charge of air that bypassed the compressor piston as the piston was nearing its maximum travel in its bore. The compressed air entered the crankcase via a bypass tube similar in operation to the way a two-stroke engine transfers gases from piston chamber to crankcase.

Some Stirling engines use pressurized gas, usually helium, and this in itself improves the performance versus air because helium is easier to push around inside a cylinder by virtue of its lighter weight. It also has improved heat-transfer properties. However, it becomes increasingly difficult to maintain an effective seal when the molecular size reduces, as it does with helium.

COOLING METHODS

There are two main cooling methods normally used in Stirling engines – liquid cooling and air cooling. In either method if the flow is increased then so will the relative cooling effect.

Both methods of cooling rely on the amount of surface area exposed to the cooling medium, be it air or liquid. For liquid cooling, a multi-channelled layout would be used; for air cooling, a multi-finned arrangement. By far the most efficient method is liquid cooling, because of the amount of energy in the form of heat that water can absorb relative to air. To put this into perspective, it takes forty-eight hours to defrost a typical frozen turkey in air, but takes half that time if defrosted in water.

When using air to cool the cold end of the displacer cylinder, a passive system might involve a large heat sink that radiates the heat into the air. The greater the number of fins and the thinner the fins, the more effective this will be.

On a water-cooled Stirling engine a reservoir of water may be used and the water can either use a simple gravity-type system or be pumped to induce the flow of the cooling medium. If the Stirling engine is for use in a boat, then the water it floats on can be used as a source of cooling fluid. It is possible to use both air and water cooling at the same time by using a radiator combined with a fan cooling the fluid in the radiator.

The greater the difference in temperature between the hot and the cold end, the more effective the Stirling engine is. It is therefore worth making the effort to pay plenty of attention to the cooling of one end of the displacer cylinder.

The length of the displacer is important because there must be sufficient length to prevent the heat from the hot end heating up the cold end, thus reducing the temperature difference. Another method that can be used to reduce the heat transfer between the two ends of the cylinder is the thickness of the material that the cylinder is made from. The thinner the material, the less metal there is to transfer the heat and it is mostly just a simple function of the cross-sectional area of the cylinder that controls the speed of heat transfer.

HEATING METHODS

The diversity of the fuel that a Stirling engine can use means that its selection can be made using almost any criteria desired. The fuel is usually used to increase the temperature of the hot end of the displacer cylinder. The methods used to make the most of the heat that is applied will determine how well the Stirling engine runs. It is worth recirculating the heat that is applied to the hot end as this improves the efficiency of the engine. The use of a material that retains heat in the hot end chamber is another method of increasing the efficiency; this is a kind of regeneration but not strictly in the same terms as described earlier. The use of a suitably heatproof insulation around the hot end and the burner assembly will radically improve the efficiency.

MATERIAL SELECTION

Materials are measured in several different ways and these different types of measurements help to determine which is the optimum material to use for a particular application. If strength is an important criteria for a component then it should be determined whether the part made from a particular material is going to be strong enough. If temperature is an issue, then finding out at what temperature the material becomes unstable is important, as is knowing its recommended service temperature. Should that high temperature part have a small clearance then its elongation in proportion to the temperature rise may become critical. If a component is required to rub against another then its relative hardness should be considered.

Components that reciprocate should be both rigid and light with enough strength to withstand the compression, tension, torsion and thermal loads that will occur. Items such as bearings need to be hard-wearing and rigid. Flywheel rims should be heavy and rigid and the spokes or disc should be light but strong enough to hold the heavy rim. Materials used for dissipating heat need to have good thermal conductivity and to remain stable in shape at the elevated temperature at which they are expected to work. Parts that rotate should be as light and ductile as possible and not be brittle.

The density of a material will be given in grams per cubic centimetre or pounds per cubic inch or other similar combinations and this gives an indication of its weight or mass in proportion to water where water is equal to one. For example, aluminium is approximately 2.7 to 2.8 times heavier than water and has a density of 2.7 to 2.8 grams per cubic centimetre. If the density of a material is less than 1.0 then it will float.

There is a wealth of data available for all the various materials that might be used in a Stirling engine and all of this data has been found through testing. The properties of these materials are in the various property categories that assist in determining the suitability of the material required (*see* Appendix II).

Many people often rely on their past experience of working with metals and other materials to gauge whether a material will be suitable for its purpose – for example, if a material has previously been used for a similar application, they judge that it will be suitable for a new component. However, while this approach has been used successfully for many years, it does have its drawbacks. Not only must

adequate experience have been built up, but also the information must be correct and relevant to the exact application. Sometimes model engineers will use scrap metal, which has the added problem of having an unknown origin or history. The cold working that is prevalent in industry today will use up a lot of the ductility that the metal had in the first place, as will the bending and cold working used in forming sheet and strip, which may be the cause of unexpected failures. The solution is to use a heat treatment such as annealing, normalizing or tempering, which, if done properly, will alleviate the problem.

In model engineering, particularly with hot air engines, the stresses and strains involved are relatively low and empirical testing and evolution of a design through failure analysis is commonplace. However, it is worth spending some time gaining further knowledge of stresses and strains, and changes due to temperature and their subsequent analysis. Some useful texts on this subject are noted in the Bibliography.

DESIGN

Before any build of an engine takes place it is important to consider how appropriate it is for the desired application. As always, the maxim, 'keep it simple', should be in the forefront of your mind. Most engineers at some point become focused on solving an immediate detail problem and do not take a step back to consider the overall aim. A good example of this is the balancing mechanism that was designed and implemented on my 10in Low Δ T, details of which can be found in Chapter 6. Whilst this is very successful, a simpler and perhaps a more elegant method is to make a hole in the flywheel and simply attach a weight as a counterbalance to compensate for the weight of the piston and the con-rod assembly. This weight will typically be on the opposite side of the flywheel and will improve the smooth running of the engine. Bob Atkins uses this method of balancing on his excellent engines, as do I on my Low Δ T Ringbom. A relative increase in performance is achieved because there is less mass involved in getting the balance of this engine correct. The reduction in mass makes the engine easier to turn and more efficient as the moment of inertia is reduced. This is only one example of simplifying a design by looking at the broader picture and this type of approach, if carried through to the design of other components, may well improve your engines.

Various methods are commonly used to assist in solving problems like this. One of the best is to take a break; have a cup of tea, walk the dog, sleep on it. Another method is to get some comments from other people whose opinion you respect, as they are often able to see directly what you have missed. However, this approach, while good, has the drawback that another individual may dismiss your ideas too readily. This may be because of their own specialism – they may have their favourite ways of doing things and may also have a less than complete understanding of exactly what you are trying to do.

Brainstorming is a method that is commonly used to assist in the provision of a multitude of ideas. It can be done in groups or individually; the idea here is to dismiss nothing out of hand, however silly it may initially appear. Often the best designs come from the seed of an idea that was initially rejected for being too far-fetched. Ensure you have many sheets of paper and consider any approach to solving the design problem and try to let your mind have free reign over a broad range of subjects, even if they are not directly related to engineering.

While taking a break it is worth reading around the subject of your design. This often gives strong pointers in ways of achieving a better design that may not have been considered; again, it is not essential to limit the choice of research to the specific subject of the design. There may be an equally valid alternative design solution used in an automotive, marine or other application that may well prove to be directly relevant to the improvement of the performance of your engine.

Look at what you have done before, question how well it worked and whether the material was suitable. Look at the way this problem has been solved historically and give more than a passing glance at the way your contemporaries have solved the problem. Do some research into new methods that are under consideration and identify if these ideas will be of benefit in your application. Nature is a good source of ideas. Look how similar wood is to some of the composite structures. Insects have a whole armoury of ingenious solutions to problems. Leaves are designed to shed and hold water yet can still be used for breathing. They provide energy and act as waste disposal. A simple walk in the woods can often bring inspiration beyond its face value.

FRICTION REDUCTION

At an engineering exhibition in Columbus, Ohio, a high-performance internal combustion engine was connected to an electric generator and an electric motor through a dynamometer. The engine was started and run at maximum power, with a power reading of 850bhp. The engine was stopped and the spark plugs were removed to eliminate the effect of compression. The engine was then run, using an electric motor, to the same rpm that it was running at when at maximum power. The measurement was then taken to see how much power it took simply to rotate the engine at that speed. This measurement was 250bhp, most of which can be attributed to friction. In effect, the total power of this engine was 1,100bhp, with 250bhp available to increase performance by concentrating on the reduction of friction.

There are various methods by which friction can be reduced. Some of them will be considered in the following pages.

Choosing Low Friction Materials
There are many types of bearings to consider. The following is a selection of the ones I consider to be most useful.

PLAIN BEARINGS

Plain bearings such as crankshaft, thrust and pin location may be axial or sliding and are not required to cope with large forces. The bearing material is mostly softer than the shaft. These types of bearings can be adversely affected by wear, mechanical stress and thermal loading:

- lead bronze is soft, does not seize readily but wears;
- brass is not suitable for high temperatures because of its zinc content but can be used for low loads;
- red brass has a higher tin content and is used for medium loads such as general sliding bearings, piston pins, bushes and crankshaft bearings;
- aluminium bronze has a high thermal expansion, which makes it suitable for interference fit bearings; it also has a reasonable wear resistance but creates more friction.

SINTERED BEARINGS

Sintered bearings are porous and are often impregnated with lubricants. They have reasonable wear characteristics and are inexpensive. They will generally run quieter than plain bearings. They can be made from iron, steel and bronze.

DRY BEARINGS

Dry bearings are made from a variety of materials. The ones I usually use are PTFE and graphite. Carbon/graphite bearings are useful in high temperature applications, and are self-lubricating but brittle.

ROLLING BEARINGS

Rolling bearings have very low friction characteristics, do not require frequent maintenance and are useful in applications where precision is wanted. They can be thinner than their plain bearing counterparts and can withstand higher temperatures, but are more likely to suffer from problems associated with the build-up of dirt. There are rolling bearing derivatives that can act as a sliding bearing, although they are difficult to seal.

Using the Correct Tolerances and Clearances

It is important to match the correct tolerance and thereby achieve a suitable clearance for the application. It is worth investing in a set of charts or tables that enable you to see and understand the types of tolerance and clearance used in industry. Typically start with BS 4500, which shows all the recognized tolerances for limits and fits for shafts and holes. These charts use microns and although it would be difficult to machine to the level of tolerance the charts indicate, often the values required for these limits and fits for shafts and holes fall within a maximum and minimum tolerance range to achieve the specified fit.

However, while these standards are a useful guide, not all of us may wish to spend so much time on this aspect. A simpler method and one that is widely used is to specify the size of the hole to a particular dimension, drill slightly undersize, then ream to the actual size. The more accurate your work is the less misalignment there will be, which in turn will ensure that the friction due to misalignment is minimized. If by using mechanical or electronic aids such as digital read-outs, indexing heads and jigs your work can be made more accurate and reliable than by using hacksaw and filing techniques, then the finished components will be better made, last longer and improve the performance of your engines.

Careful Machining, Polishing and Finishing of Components

By taking the example of a machining a shaft, cylinder or a piston, the more care exercised in the machining, honing, lapping and polishing, the less friction, binding and seizing there will be. The finer the finish on a shaft or piston, the smoother it is in operation. The time invested in the finish pays dividends in the quieter and more efficient operation of an engine. A method of achieving a finer finish when approaching the final size of a machining operation is to ensure that progressively finer cuts are taken. This lessens the surface stress of the part and reduces the possibility of cracks and misalignment that will cause friction in use.

Use of Appropriate Oils and Other Anti-Friction Agents

The role of a lubricant is to keep two components separated while they are moving against each other. Lubricants can be gas, liquid or solid and are chosen with regard to the load, the materials of the two bearing surfaces, the design requirements and the conditions in which the components are expected to operate. A secondary function of a lubricant may be to act as a seal, a coolant or to inhibit corrosion or noise. Additives are a useful means by which to influence the fluidity of the lubricant at various temperatures or pressures. The viscosity of a lubricant should be appropriate to the application for which it is destined. The nature of Low Δ T engines often means that lubricants have to be removed as far as possible with only the merest hint remaining to inhibit corrosion. For medium-power Stirling engines low viscosity automotive engine oils are acceptable, although sewing machine oil or other lighter oils are preferable for models. Additives such as Teflon, silicon or graphite can be useful and are worth trying.

When thinking about additives, oil companies spend millions refining and developing a formula that is just right. It may therefore be a little presumptuous to think that you know better and that, by adding a little of this or that, you can improve on what they have done. This is not quite so important in non-critical applications, but certainly noteworthy.

Coating or Plating of Surfaces

Various plating methods are available depending on the material being used. Chrome plating for steel and anodizing for aluminium alloys give a lower friction coating that transforms the surface hardness of a component. If adjacent parts, such as shafts, gears and linkages, which are made from copper, brass or other materials that readily work-harden bear on each other, they will, by the nature of the material they are made from, gradually and progressively work-harden each other. However, if these adjacent parts consist of the same material, there is no need to coat or plate.

DEAD SPACE REDUCTION

Dead space is the volume available for the expansion of the working media before they start to act on moving the power piston. The reduction in dead space improves the performance of a Stirling engine. Ways to reduce the dead space are to:

- reduce the ID (Inside Diameter) or cross-sectional area of the pipe or slot that connects the displacer cylinder to the power cylinder;
- reduce the gap between the pistons and the end of their respective cylinders;
- place the displacer piston/cylinder as close as possible to the power cylinder. (This technique is maximized in the Beta-type engines where the power and displacer pistons are in the same cylinder.)

If using a regenerator, maximize the internal volume space with the regeneration material up to the point where you don't impair the flow.

ALTERNATIVE OPERATING MEDIA

Some of the media that can be used inside a Stirling engine are air (nitrogen and oxygen), helium and hydrogen. There are also arguments for using inert gases such as argon and freon. If using gases other than air, careful attention should be paid to the design of the heat exchangers and to the thermal conductivity and the specific heat capacity of the chosen gas and of the material of the heat exchanger.

Solo Stirling 161, a combined heat and power unit shown open.

Microgen combined heat and power unit. (Reproduced with the kind permission of Microgen)

FUEL

Probably the major factor that will ensure an exponential growth in the mainstream use of Stirling engines will be the need to maximize the use of our existing energy reserves such as oil and gas. The basic efficiency of the Stirling engine approaches nearly twice that of an IC engine and can be up to 15 per cent more efficient than a jet or CI (Compression Ignition – diesel) engine. In the short term this will return a 25 per cent fuel saving in areas such as CHP (Combined Heat and Power), where the energy used to heat domestic hot water can be used to generate electricity instead of wasted. Any excess electricity can be returned to the National Grid and as such the Stirling engine is placed to meet policies to decentralize power generation. As our reliance on fossil fuels diminishes, the ability of the Stirling engine to run efficiently on any fuel that is capable of producing heat (or relative cold) such as methane, hydrogen (fuel cell or otherwise) or

renewable fuels such as wood chippings or straw will go a long way to ensuring its continued development and success.

Methods utilizing natural and renewable energy sources that are pertinent to the advantageous use of the Stirling principle include power generation from solar energy in either a simple focused dish engine or an auto-tracking multi-faceted mirror array generator for power stations. Geothermal sources are also ideally suited for use with this type of engine.

Most model engines use fuels such as methylated spirit and bottled butane/propane. Some low-power examples use nightlight-type candles, and Low Δ T engines can use excess heat from 40W light bulbs and even body heat. Remember never to use flammable liquids anywhere near a naked flame, ignition source or hot engine and only decant the amount required. When refuelling burners, refuel in a well-ventilated area, preferably outside. This will minimize any chance of the vapour

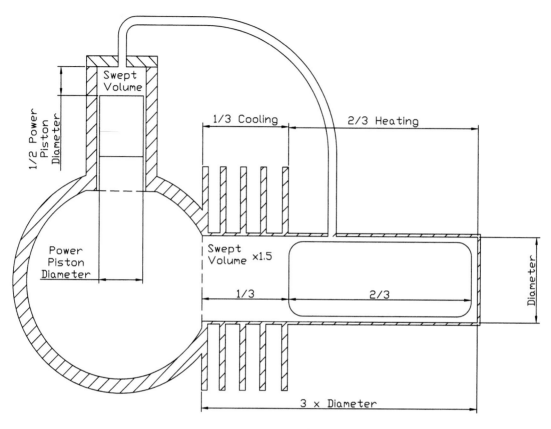

Relative sizing recommendations for a Stirling engine.

catching alight. Always use an appropriate sealed container to store fuel and use a suitably sized tube or funnel while refuelling to prevent spillage. If a spillage does occur, mop it up immediately. Do not light any engines that have had fuel spilt on them until they have been cleaned and always ensure that burner vent holes have been checked and remain clear.

3 Design Variations

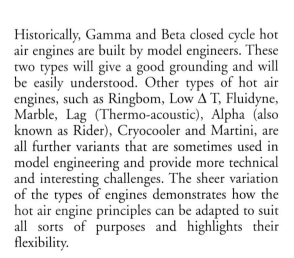

A simplified representation of a Gamma engine.

A simplified representation of a Beta engine.

Historically, Gamma and Beta closed cycle hot air engines are built by model engineers. These two types will give a good grounding and will be easily understood. Other types of hot air engines, such as Ringbom, Low Δ T, Fluidyne, Marble, Lag (Thermo-acoustic), Alpha (also known as Rider), Cryocooler and Martini, are all further variants that are sometimes used in model engineering and provide more technical and interesting challenges. The sheer variation of the types of engines demonstrates how the hot air engine principles can be adapted to suit all sorts of purposes and highlights their flexibility.

GAMMA

The majority of model engines are of the Gamma configuration. This type has two separate cylinders, a displacer cylinder and a power cylinzder, joined together by a pipe or slot in the casting. The two cylinders can be bolted to a base or frame and can be connected to a single crank and flywheel. It is a very simple engine and ideal for model engineers making their first engine. The disadvantage of this type of engine is that the connecting pipe is dead space, which leads to inefficiency. However, this should not worry the model engineer too much on their first engines.

BETA

The Beta engine, sometimes referred to as concentric, is the most efficient. One of Rev. Robert Stirling's first engines was a Beta engine. The displacer piston and power piston are both in the same cylinder. The displacer piston is operated by the displacer rod, which runs through a gland or bush right through the middle of the power piston. The displacer piston runs at the hot cap end of the cylinder and the power piston at or beyond the cold end.

There are two main reasons for its efficiency. Firstly, there is very little dead space since no connecting pipe is needed as with separate cylinders. Secondly, at one stage of the cycle the two pistons come very close together. In Stirling Silver II and the Solar engine the two pistons come to within a few thou of each other. This causes a lot of turbulence, which is one of the things that makes the Stirling cycle work well; the greater the turbulence the better the results.

A connecting rod to a crankpin is needed to drive each piston. To get symmetrical thrust on the power piston, two con-rods are required and pass either side of the displacer con-rod. This extra linkage in a Beta engine causes more friction, but the overall efficiency of the engine more than compensates for this. Some Beta engines use one rod for the power piston, but this is not ideal as it puts unequal pressure on the piston and causes more wear and more friction.

There is a disadvantage when running a Beta engine horizontally due to the load of the displacer bearing down through the power piston guide. This transfers to the walls of the cylinder via the power piston, creating drag. Despite this, it is still very efficient.

One solution is to run the engine vertically, so that the weight of the displacer transfers on to the crankpin and the problem of drag is not so marked. In the vertical position a ring-type burner is required. Ring burners are not easy for a model engineer to make and are rarely as successful as a burner that is used under a horizontal cylinder. The correct length of flame between the burner and the displacer cylinder is important. A nice blue flame is the optimum flame to aim for.

Usually the cylinder housing the displacer, and the power pistons, are the same internal bore throughout their length. In a lot of other engines the power piston is a smaller diameter with a longer stroke to achieve the capacity or required swept volume. In a Beta engine the stroke of the power piston is shorter to accommodate the large bore relative to the swept volume. A shorter stroke means that the engine will complete the cycle quicker and consequently will rev more easily.

Linear engines are Beta-type engines. They have no linkage. The displacer and power piston are suspended by springs and the pressure variation operates the engine. The balance of the springs needs care to set up. A linear alternator is connected to the piston, which produces alternating current.

ALPHA/RIDER

Named after A. K. Rider, who first manufactured this type of engine, this configuration has two separate cylinders. One is the hot cylinder and the other the cold cylinder. There is a power piston in the cold side. In the hot side there is also a power piston, which has a displacer mounted on to the crown of it. The two pistons have an adjoining tube. Cooling fins will usually be found on the cold cylinder.

One advantage is that the engine lends itself to regeneration. The two cylinders can be mounted in a 'V' at 90 degrees to each other. Each con-rod is coupled to a single crankshaft. The connecting pipe, although dead space, can be filled with regenerative material such as stainless steel gauze or ceramic beads, creating an external regenerator built between the two cylinders. The dead space can be successfully reduced by using this technique.

The Alpha can make a compact four-cylinder engine in which each piston has a

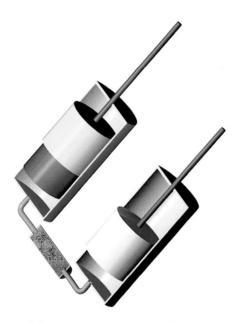

A simplified representation of an Alpha engine.

principle and the diaphragm moves with the reciprocation set up by the heat differential. There are very few moving parts in the engine and subsequently it is a design that has the potential to be reliable in service.

FLUIDYNE

A Fluidyne engine works on the Stirling principle whereby a heat source heats up one end of a chamber that is cooled at the opposite end. This air chamber takes the form of a tube. The pressure change that takes place in this chamber acts at either end of the tube on a

double use. The top of the piston is a displacer and the bottom is a power piston. Each piston has a 90-degree advance phase to the next one. The top of one piston is connected through a regenerator to the bottom of the next piston.

This system can be used on in-line or square engines. On a square engine, parallel con-rods are used. On the square engine an oscillating disc is used, which is similar in operation to a swash-plate. This is known as a Siemens square engine, a design patented by Sir William Siemens in the mid-1800s. It is double acting in as much as one cylinder's pressure is entering the bottom of the next cylinder in a square.

THERMO MECHANICAL GENERATOR

Components that are typically used in a Thermo Mechanical Generator (TMG) include a diaphragm spring that is attached to the displacer piston and its output is usually through a linear generator. It uses the Stirling

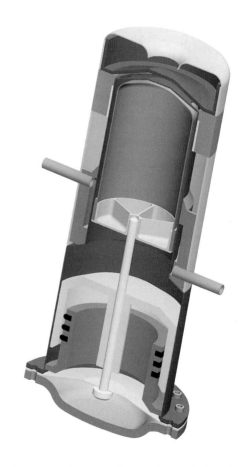

This free piston Stirling engine is similar in function and layout to a TMG.

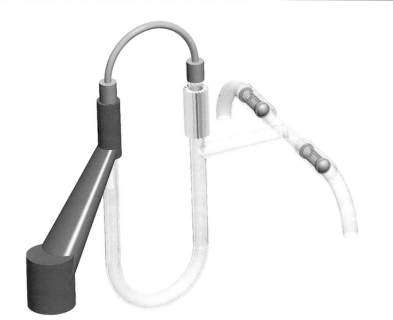

A typical fluidyne layout.

Roy's first fluidyne engine utilizing a kilner preserving jar, some copper tubing, one-way valve assembly and a beer glass. (Reproduced with the kind permission of Graham Darlington)

Roy demonstrating a fluidyne engine to the Duke of Gloucester. (Reproduced with the kind permission of Mike Chrisp)

piston. That piston is in fact a fluid and is generally in a 'U'-shaped tube which is connected to either end of the underside of the air chamber (tube). This fluid piston moves up and down as the pressure of the air above it changes. Teed off from the cold side of the fluid piston is a further tube which connects to another tube via a second tee. This tube has two one-way valves in series on opposite sides of the second tee. Both valves flow in the same direction. As the air pressure increases it forces part of the fluid that makes up the piston out through its only escape route, one of the one-way valves. When the pressure decreases it draws fluid in through the other one-way valve. This completes the cycle. This cycle is similar to the cycle found in thermo-acoustic engines.

Fluidyne engines are self-starting, and the only mechanical moving parts are the one-way valves. The one-way valves can be made from off the shelf components or can be adapted from other applications such as reed valves, tyre valves or one-way air valves used for fish tanks. The fluidyne engine is comparatively inefficient with regard to Stirling engines in general, but its simplicity and niche applications such as water pumping from waste heat outweigh this.

SINGLE/DOUBLE ACTING

A double-acting design is appropriate for a four-cylinder Stirling engine. The reason for this is that the power stroke must be 90 degrees out of phase with the displacer piston's stroke. If there were just two pistons it would be 180 degrees out of phase and would be unlikely to work. The power and the displacer piston in this instance are the same component and have working gas above and below. The working gas on the underside of the piston will work with the next piston in sequence and the working gas above the piston will work with the previous piston. The gas from the underside acts on the top side of the next piston in series. In effect, the top of the piston chamber is the hot space and the underside is the cold space, although the reciprocation set up by the heat differential

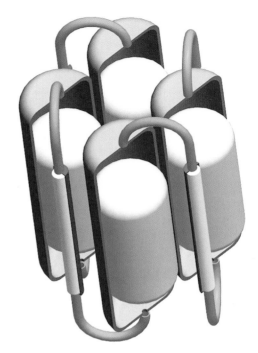

A simplified view to show a double-acting layout.

acts on a different piston in the adjacent cylinder. This double-acting design is very economical with respect to the amount of parts required to build the engine and is typically used for high-power engines and engines that are proposed for volume manufacture. One of the most well-known examples of this engine layout was used in a Ford 'Torino' which was built in a joint venture with Philips.

This book mainly concentrates on single-acting Stirling engines because the operation and the engineering are more straightforward and easier to understand. Single acting usually means that there is a separate displacer and power chamber, except in the Beta type where the cycle takes place in one set of piston chambers and acts on one power piston.

DRIVE MECHANISMS

Drive mechanisms are used to ensure that the drive from the pistons is maintained in a given

A crank lever assembly.

axis. It is important to keep the motion of pistons as true as possible, as doing so will help to eliminate premature wear. The methods described are the ones I usually use. There are many variations on this theme and the ones below are simple, work well and reliably, and are appropriate for model Stirling demonstration engines.

Crank and Lever

A crank and lever is a very simple way of altering reciprocating motion and turning it into rotary motion or vice versa.

Scotch Crank

The scotch crank method of converting reciprocating motion to rotary motion is to restrict the movement of the con-rod to the fore and aft on the piston axis. This simplifies the drive, as it eliminates the requirement for a con-rod and clevis assembly because there is no angular movement of the con-rod. The scotch crank can be either a slot or wire shaped as a guide that is similar in appearance to the key used to open a corned beef tin. The slot or guide allows the crankpin to move up and down as it rotates without moving the con-rod vertically. It moves the con-rod along the piston

This view shows a scotch crank method of converting reciprocating motion into rotary motion.

axis as the crankpin rotates on the crank disc. This pin bears on the edges of the slot or guide of the scotch crank assembly.

GEAR DRIVE

A gear drive is used where smooth or powerful operation is necessary. It comprises of a pair or

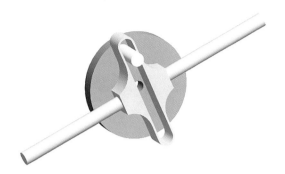

A crank disc, pin and scotch crank assembly.

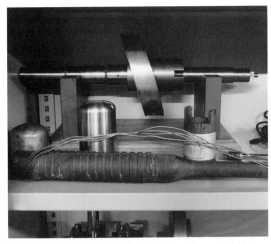

This view of a swash plate was used in a Philips engine was taken at a private collection in Holland.

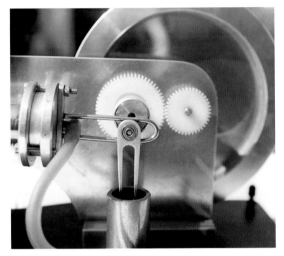

This shows an offset gear drive that connects to a flywheel and drives the displacer via a scotch crank.

more of meshing gearwheels that spin a flywheel faster to store more kinetic energy for a given flywheel size factored by whatever the ratio of the gears are. It may also be used to transform the path of the drive through 90 degrees with a crown wheel and pinion gear assembly.

Swash Plate
A swash plate is a circular plate that is connected to a shaft at an angle. The shaft attaches or goes through the plate along the centre axis. As the shaft rotates the angled disc pushes a con-rod that bears on the disc surface

in and out. This causes a parallel motion of a piston con-rod to the shaft axis. Despite appearances, this drive is very smooth and well balanced. It can be used for both the displacer and the power pistons and has the advantage that many pistons can be arranged to work radially and parallel to the shaft axis. The stroke of the piston is controlled by the inclination of the disc to the axis of the shaft and by the diameter of the swash plate.

Ross Linkage
This linkage was developed by Andy Ross, who has designed some very good engines, and the way the Ross linkage turns circular motion into reciprocating motion is quite impressive. Even though it looks complex, it is in fact fairly simple. It is far more robust than many other drive mechanisms and I tend to use it on engines that are more powerful, the exception being a simple Ross linkage on one of my early Low Δ T engines. It operates with a swing arm that pivots on the end of the con-rod. This pivot is positioned at both extremities of the tee or triangle and ensures that the con-rod maintains its axial position relative to the piston travel. The centre pivot on the tee piece or triangle connects to the crank disc.

Above *A Ross linkage, showing a method of turning reciprocating motion into shaft power.*

Another Ross linkage on a Gamma engine, showing the power piston con-rod at the bottom and the displacer piston con-rod at the top.

4 Gamma Hot Air Engines

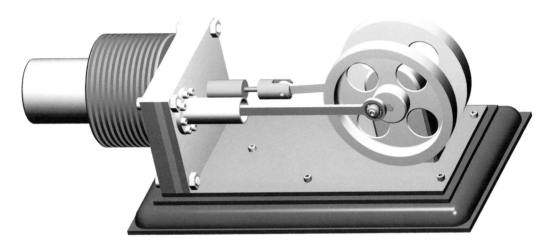

A simple Gamma view with twin flywheels.

A typical Gamma engine has two pistons, one power piston and one displacer piston. These two pistons are usually coupled to a common crankshaft with a crank disc and a flywheel. Where the displacer piston is connected to the flywheel it is typically attached 90 degrees in advance of the power piston attachment. This may be varied by a few degrees advance or retard, but 90 degrees is easy and works well. There has to be a passage for the air or gas to move from one cylinder chamber to the other. Gamma engines can be air-cooled or water-cooled and examples of both will be found in this chapter.

> The Gamma engine is the simplest form of Stirling engine and is the one I would recommend to everyone as their first engine.

GAMMA MK I

This engine is based on James Rizzo's 'Dolly' engine.

The power piston is made of cast iron, running in a mild steel cylinder. Experience and wear patterns show that the cylinder would be more effective if made from cast iron, as the self-lubricating qualities of cast iron in cast iron are superior to cast iron in mild steel.

The main block that the cylinders are bolted to is made up of two $^1/_4$in plates and one $^1/_8$in plate. The $^1/_4$in plate is sandwiched between them with a paper gasket either side and a slot is cut in the $^1/_8$in plate, which acts as the connecting tube between the displacer piston and the power piston.

Gamma Mk VI.

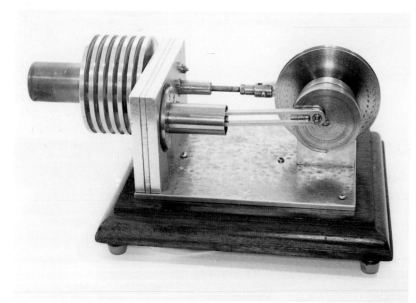

Gamma Mk I. This gives a view of the power cylinder, the flywheel, the power piston con-rod and the sandwich construction of the transfer chamber.

The displacer cylinder has a very thin wall, about 0.020in thick. It is made from stainless steel, as is the displacer piston, which has a clearance of about 1/32in all round. Fitted to the piston is an alloy blanking-off cap that is bonded on to the cold end with Araldite glue. Superior results are achieved when bonding if the quick-setting Araldite is avoided.

The displacer piston is on the end of a silver-steel rod that runs in a nicely fitted brass bush.

Brass and silver steel will work well together, but phosphor bronze instead of brass would last a lot longer. It is desirable for the main shaft to run in ball races and it should have ball races on the big ends of the cranks.

The whole engine is bolted to an aluminium base, which is in turn screwed to a wooden plinth. The underside has a magnetic mount to hold the engine steady and in position when displayed on a steel tray at shows. Both

This gives a good view of the cooling fins surrounding one end of the displacer cylinder and shows the hot end, sometimes known as the 'hot cap'.

In the foreground is the power piston assembly showing the power cylinder and its attachments, the lightweight con-rod and the crank disc, which attaches to the con-rod using a steel crankpin.

cylinders are flanged and are held to the main cylinder block with 6BA screws. The screws for the cylinders also keep the plates together.

The seven cooling fins are made from aluminium alloy and were turned from a solid block. The size (surface area) of the cooling fins prevents this engine from overheating. The crank pins for the little ends are made from silver steel. This engine has been running and displayed at many shows over the last fourteen years and during that time has needed very little attention.

> I must emphasize that when making any hot air engine, it must be gas/airtight, and a lot of attention must be paid to keeping friction to the minimum. Stirling engines will not withstand friction or air leaks of any kind.

GAMMA MK II

The cylinder is of steel, and flanged top and bottom with a power piston running inside it. The power cylinder is at 90 degrees to the displacer.

In this version, the displacer cylinder and piston are a separate unit from the rest of the engine. It has a very large stainless-steel displacer with large 6in diamter cooling fins. These fins were originally hard disk drives from an early computer. They are bolted together with three screws, with a washer in-between each disk to give the spacing.

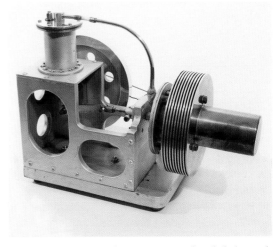

Gamma Mk II with ex-computer hard disks and hot end.

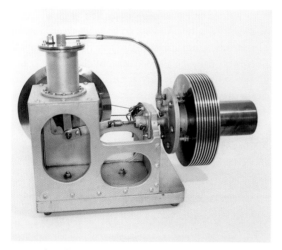

Note that both con-rods use the same crankpin.

View of the flywheel connected to a generator and the power cylinder and transfer pipe.

Conveniently, a single crankpin is used to drive the displacer piston and the power piston via their con-rods. The top of the cylinder plate has a hexagonal brass plug screwed into the top of the cylinder. This can be removed to put in a few drops of oil for lubrication, which enable it to run for several days at a show. The crankshaft is counterbalanced to offset the weight of a large piston, thus making this a very smooth-running engine. The external pipe that connects the two cylinders together has been kept as short as possible.

The 8in diameter flywheel drives a small 12V generator. The two aluminium bearing housings are fitted with the main bearing ball races. The cylinder is bolted to a square aluminium plate that in turn is bolted to a wooden frame, which is painted silver. This gives the appearance of an aluminium casting.

This engine has now been rebuilt. It was made in my early days of modelling hot air engines and the displacer cylinder was made from a tin can. A salmon tin was used as a water jacket and was soft-soldered on to the main tin of the displacer cylinder. This engine ran for a few years. The displacer tin (the inside of the water jacket) eventually rusted and allowed a few drops of water inside the displacer cylinder. I used a blowlamp on the end and the heat turned the water into flash steam. It ripped all

the solder joints of the can, there was a terrific bang that could be heard from quite a distance away and bits of the engine were found on several of my work cabinets some feet away. This was a warning that if steel is used and the engine is water-cooled it will rust in a few years and if any water finds its way into the displacer cylinder it can be disastrous. Now, I would only recommend the use of stainless steel for a displacer cylinder. When I had rebuilt the engine, people asked what had happened to that lovely old flywheel because it was varnished oak and with the salmon tin and a cat food tin it looked quite intriguing. During the rebuild, the engine was modified to become an air-cooled engine.

GAMMA MK III

This engine is my version of James Rizzo's 'Dolly II'.

The cylinder layout is similar to the 'Dolly' engine, in which the power and displacer cylinders attach to a cylinder block. The cylinder block is mounted on a 1/8in brass plate. Although this engine uses castings, my

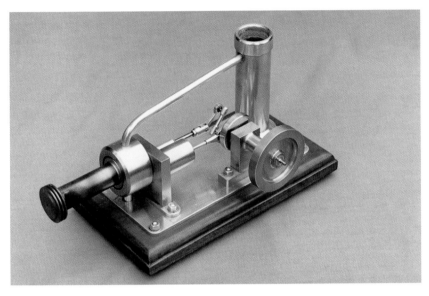

This shows a good view of the entire Gamma Mk III. In the foreground is the brass flywheel and the steel displacer cylinder. Note the fins on the hot cap. (Reproduced with the kind permission of Graham Darlington)

preference is usually to machine from solid, as castings tend to be difficult to hold and their patterns have to be made. When it comes to the cylinder you need to be able to lap the cylinder right through. It is not possible to lap the blind cylinder effectively, so I use a loose-fitting steel liner which is then fixed in place using Loctite. In this engine the cold end of the displacer and the power cylinder are part of this casting. The displacer cylinder is made from mild steel tube that is slid into the casting on two O-rings. This steel tube had an end brazed on it and was

turned to form fins on the end. There is not enough heat generated to melt the brazed end, but the same would not be true if silver solder was used. Silver solder will withstand a methylated spirit flame but not an intense gas flame. If using only methylated spirit, silver solder is quite good enough.

Ball races are let into the brass main bearing housing. There is also a ball race in the end of the displacer drive The big end on the power cylinder con-rod and displacer con-rod also have ball races in them. The main bearing casting was supplied by James Rizzo. I utilize bits and pieces from all kinds of things and the cooling tower was an old refuelling nozzle assembly that refuelled Lancasters during my days in the RAF.

This engine has a thermo-siphon water-cooling system. The hot water goes from the top of the water jacket up an angled tube to the top of the water tower. This tube must not be parallel because hot water will want to rise. The reservoir of water in the tower slowly cools and the cool water sinks to the bottom of the tower.

Gamma Mk III water transfer tube and cooling tower. In the centre is the piston linkage.

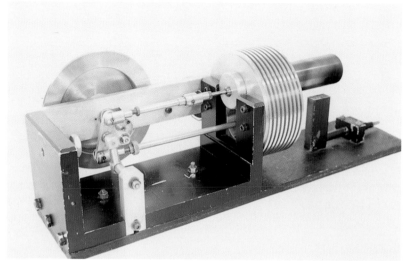

Gamma Mk IV.

It returns to the underside of the water jacket via a pipe underneath the engine; this then completes the cooling circuit. Water cooling is the main difference between this and the other Gamma engines that I have built. No pumping is required as it is designed to use the thermosiphon principle. The water jacket is sealed and mounted with two O-rings and does not require any other fixings. This has successfully sealed the water jacket for over ten years.

GAMMA MK IV

This is a Gamma engine in which the displacer and the power cylinder are all machined into the same block. A large piece of aluminium is used for the block and the gasket between the block and the cylinder head has a cut-out that acts as the transfer slot.

This engine uses the Ross linkage, having a swing arm. Originally it was built using a central guide for the Ross linkage and there was an arm from the centre pin coupled to a clevis. The guide rod that keeps everything parallel was planned to run there. However, to achieve this a support was needed and the load was too great for it to run free. I reverted to the original Andy Ross swing arm. It is possible to use a parallel guide as is demonstrated in Stirling Silver I (*see* Chapter 10).

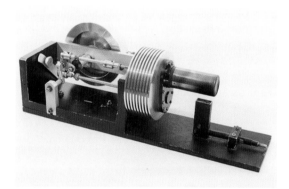

A good view of the adjustable con-rods attached to the Ross linkage.

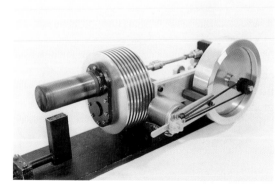

Note the method of attaching the flywheel to the shaft.

The con-rods are made from aluminium tubes. The linkage uses silver-steel crankpins that run in phosphor bronze bushes. The main bearing housing has ball races at both ends and is flanged to keep it square when screwed to the frame.

The countersunk flywheel is held on with a split collet and a countersunk nut. The boss is threaded and the gland nut goes over it. An O-ring is slid over the shaft and when the countersunk nut is tightened, it squeezes the O-ring and it clamps the flywheel onto the shaft without marking it. I prefer this method as it is easier to make and the friction grip is quite sufficient.

GAMMA MK V

This gas-fired and air-cooled engine was built as an experimental engine. The cunning plan was to build an engine without lubrication and to see how long graphite would last. The engine therefore had a graphite power piston, a graphite guide bush for the displacer rod to run in and in the scotch crank was a graphite bobbin.

The first thing to fail was the bush that guides the displacer rod. It was replaced with phosphor bronze. The second thing to fail was the graphite bobbin. A phosphor bronze bobbin was substituted. The graphite power piston, which powdered up and had to be removed and cleaned regularly, was the last to show signs of wear and was replaced by an anodized aluminium piston.

The engine as it is now has run for hundreds of hours at over 1,000rpm. Although graphite is ideal for Low Δ T work, for a high-speed engine and for highly loaded components it is not particularly suitable. There are special grades of ceramic graphite that may overcome this. The experiment of no lubrication was worth trying, and although perceived as a failure it provided some useful information which was utilized in my later engines. It is all part of a wider learning curve.

The power cylinder enters the displacer cylinder at 90 degrees. This type of engine

Gamma Mk V, consisting of a square power cylinder and a cross-drilled displacer with 90-degree power piston. In the foreground is the flywheel and generator.

always runs well. The reason is that gases are agitated much more by the turning through 90 degrees. Another reason is the action of the power piston acting directly down into the displacer cylinder, which is very similar to the way that a Beta engine works. I usually consider a Beta engine to perform best, but this type comes very close.

It is fascinating to watch with the beam flying up and down like the rocker on top of a camshaft. It was designed to be a single-cylinder version of a multi-cylinder engine. The eventual idea was to make it larger, possibly with six or eight cylinders, and to be able to operate a locomotive. Despite the initial failures while testing the idea, the engine itself was eventually very successful and if you had six in a row with all the beams going up and down, it would have the appearance of an internal combustion engine with the rocker box cover off. It certainly attracts a lot of attention.

The power piston was lapped, polished and, after the engine had been run for a while, the piston was taken out and anodized. It is a perfect fit and now there is a very hard surface on the soft aluminium piston. The power cylinder is

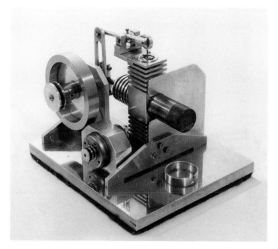

A good overall view of Gamma Mk V.

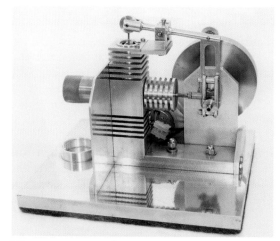

This reverse view of Gamma Mk V shows the scotch crank and reciprocating beam.

made of 1¹/4in aluminium alloy square bar and that also forms a support for the engine.

The displacer shaft operates the stainless-steel displacer piston through a guide bush that crosses the casing. There are cooling fins cut both above and below the displacer cylinder and the power piston is in the vertical section.

There are two ball races for the main bearings, and another pair are fitted to the beam that connects the power piston to the con-rod.

The power piston drives the scotch crank by means of a small beam. The pivot of the beam runs in two Oilite bushes and a ball race in the big end and the small end of the power piston con-rod. It has a brass flywheel which is 3¹/2in diameter and drives a small generator, which generates enough power to light ninety LEDs

This Gamma engine uses a ¹/2in thick base of cast aluminium. The base is 6in × 7in. Cast aluminium in sheet form is quite difficult to obtain; this came from a scrapyard. It is used in the aircraft industry for jigs and fixtures and it remains stable. It has been machined very accurately and has a wonderful finish that makes for a lovely base. There is a magnet built into the base, because several of my engines run on a large steel tray at shows and the magnet anchors them down to the tray easily and quickly.

This engine is gas-fired and air-cooled.

GAMMA MK VI

View of the complete Gamma Mk IV, showing the hot end.

This engine has always attracted much admiration due to its unique shape, futuristic appearance, compactness and smooth running. It remains one of the engines that is a stalwart of my collection and is used at most of the

The linkage and con-rod assembly of the Gamma Mk VI.

The cooling fins and cylinder block assembly from the rear, and the flywheel mounting.

shows at which I exhibit. It has no base but is constructed on three feet, one of which is a magnetic disc and the other two are suction cups. These feet secure the engine safely and firmly when being demonstrated at shows. The engine runs so smoothly and silently that no movement can be detected with the naked eye and onlookers often assume that the engine is not running.

The displacer cylinder is made from stainless steel, as is the displacer piston. The displacer piston has an aluminium plug bonded into the end of it, which connects via a brass tube to one side of the Ross linkage. The material for the swing arm of the Ross linkage was made from an oil-impregnated nylon and now uses phosphor bronze bushes. The aluminium clevises that attach the con rods to the Ross linkage are chamfered to blend into the con-rods. Attached to the other side of the Ross linkage is the power piston.

The power cylinder is aluminium in which runs a lightened leather piston impregnated with oil. This needs very little lubrication. If the power cylinder was anodized the smoothness of the finish would improve the performance by reducing friction; if anodized black it would also aid cooling. The front end of the crankcase has a cover that is held on by several brass screws and the gasket that seals this has a slot cut into it to allow the air to transfer from the power cylinder to the displacer cylinder. The cooling fins are machined around the outside of the crankcase, which is bolted directly to the frame.

5 Ringbom Hot Air Engines

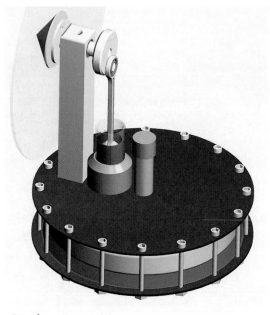

Ringbom Low Δ T type.

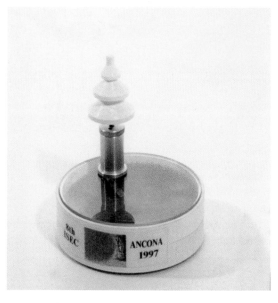

A small demonstration engine that was available from one of the early international conferences.

A Ringbom has no mechanical connection between the displacer piston and the power piston. The displacement of the power piston displaces the air, pushing the displacer piston up and down, as does the pressure reaction that takes place.

LOW Δ T RINGBOM

The displacer piston is suspended in the cylinder halfway up the stroke by a spring attached in the centre. The other end of the spring is connected to a chamber above the displacer cylinder. It has an adjustment screw in the top and there are air holes into the cylinder on both sides, so that when the pressure drops inside the displacer chamber (when the air is cold) atmospheric pressure goes into the top of the piston and helps to return it.

> Some Ringbom engines are very similar to the traditional Low Δ T engines, but without the mechanical connection between the displacer and the power piston.

48

This shows most of the main components that go to make a Ringbom engine.

BDC of the displacer piston.

TDC of the displacer piston. The tube that contains the spring from which the displacer piston is suspended is also visible, as are the nylon screws that hold the assembly together. Note the position of the power piston relative to the displacer.

The power piston going up and down powers the displacer. There is a weight on the cylinder at the front, which acts not only as a handle to pick it up, but also keeps the whole mass of the engine symmetrical. This helps to keep it balanced when operating on a frame and has the advantage of providing additional cooling.

The power piston is made of graphite running in a glass cylinder. The glass cylinder was cut from a hypodermic syringe originally used by a vet. One hypodermic was cut into three pieces and used to make three engines.

Two round plates on the top and bottom of the engine are the hot and the cold side. These are held together by nylon screws. There is O-ring material placed in a groove turned into the two plates so that the perspex displacer sidewall tube sits on this O-ring top and bottom to seal it. When the nylon screws are tightened, they help to make the displacer chamber airtight. The nylon screws stretch when tightened and any excess that protrudes is cut off.

The little Ringbom Low Δ T engine often self-starts, which can be difficult at shows because if you stop it for a demonstration it will immediately start running again.

The Ringbom engine shown here has an expanded foam displacer piston attached to a brass guide rod running in a bronze cylinder with a very light spring. The displacer piston floats or hangs on the spring and is delicately

49

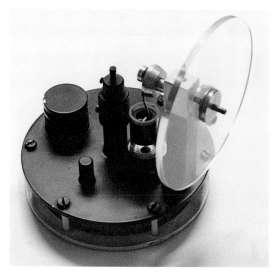

From right to left: flywheel, crank support, power piston and cylinder, spring support, displacer air bleed valve and finned weight housing.

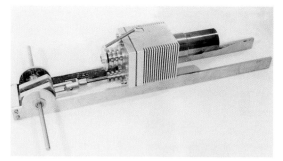

Square-finned Ringbom. A good view of the entire engine.

balanced. Its position at rest can be altered by raising or lowering it with the adjusting screw.

The main shaft rotates on ball races that are machined into both sides of a $5/8$in square aluminium bar. These bearings support the perspex flywheel, which uses a counterbalance weight on it to balance the weight of the piston. The big end of the con-rod has a tiny ball race and a plain bearing inside the piston.

> This Ringbom engine has a threaded and knurled bleed valve with which you can bleed air into the displacer, but I have never found it necessary to use this feature.

SQUARE-FINNED RINGBOM

This Ringbom engine is still under construction. The concept for this engine is to make a small buggy which will run up and down a track at model engineering shows. This is particularly suited to using one of these engines, as a Ringbom will run equally well in either direction and a mechanism that reverses the buggy when it hits a buffer at either end of the track will be fitted. It should be capable of running up and down the track all day without attention.

The main block is cut from a solid piece of aluminium. A hole has been bored out for the displacer cylinder to screw into. The power cylinder is bored into the block below the displacer cylinder. The displacer cylinder was machined from a solid bar of titanium. This machining operation took a whole day and created plenty of swarf.

> A word of warning. When machining titanium or magnesium the swarf created is highly flammable. If the temperature of the swarf is high enough it will burn with an intense white flame. These metals when molten react violently with oxygen. The only sensible method to reduce the risk of ignition is to use copious amounts of coolant during machining and a very slow cutting speed and feed. Always stop at frequent intervals and clean up the dust and swarf. If you are particularly concerned about machining these metals, consider having a professional engineering company do the machining. There are three things that cause and sustain fire. These are heat, oxygen and fuel. By removing any of these components the fire will extinguish. In circumstances such as with these metals the fire will be very difficult to put out. It would be prudent to construct a gas chamber over the cutting area and pressurize it with an inert gas such as argon. When I was stationed in Accra I was witness to a fire that was started by machining magnesium on a lathe. The fire was not controlled and despite best efforts the entire hangar burnt down with the loss of about thirty vehicles.

This shows the aluminium block and titanium displacer cylinder.

The displacer piston is made of stainless steel with an end brazed into it. The displacer piston guide rod is ³/₈in silver steel which has been bored down the middle to make a very thin-walled tube. In this Ringbom the displacer rod is hollow to allow atmospheric pressure inside the displacer. There are O-rings at each end of the displacer rod. The O-rings give a slight buffer effect and without these the metal would make quite a clatter. The guide for the displacer rod is brass and is made with four external cooling fins. The guide for the power piston is also brass with a silver-steel piston rod running inside it, and this also has four external cooling fins.

The gasket has a ⁵/₈in slot cut into it which allows free flow of air from the displacer cylinder to the power cylinder. The frame is ¹/₈in aluminium sheet. The bearings are mounted to a semicircular flange that forms part of the frame at the rear. The big end is connected to the piston by an aluminium con-rod. This couples to the crankpin, which connects the two crank discs together. The flywheels mount on to two silver-steel rods.

This engine does not yet run to my satisfaction and I am currently constructing a cast-iron piston that should prove more suitable. When optimum running is achieved, the frame will be finished and wheels attached.

FREE PISTON TEST RIG (MODIFIED RINGBOM)

Free piston test rig.

The idea behind this engine was to try to get a free piston engine running. I made it on a column, similar to a drill stand. This helped to get different heights and gave scope for experimentation. The basic design is a Ringbom with a variation on the power piston.

The displacer cylinder has the hot end at the bottom with cooling fins above. Immediately behind the displacer cylinder is the power cylinder; the air passage between the displacer and the power piston is cut into a gasket so that the air runs along the gasket to the power cylinder. This also acts as a thermal barrier. The brass power cylinder is extremely long. The aim was to create a free piston, where the power piston moved unrestrained up and down the cylinder without popping out of the end; when it reached the end of the cylinder the displacer would suck it back. This was only successful for a few minutes whilst the piston worked its way further and further up the cylinder before

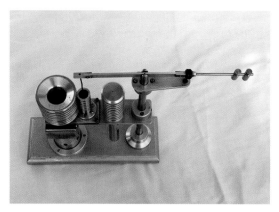

Top view of the free piston test rig. (Reproduced with the kind permission of Graham Darlington)

the piston to self-align with the natural cycle. Although the piston is coupled to the beam it is able to alter its phase to line up with the displacer. The piston follows the displacer but doesn't have exactly the same stroke as the beam. This makes it fascinating to watch.

When the engine starts, it starts slowly and reciprocates with a low frequency using maximum movement on the beam. A rubber buffer is provided for the excess movement at start up. Once the engine increases in temperature, the amplitude of the beam's oscillations is less and the rubber buffers are not used. The adjustable weights can then be moved towards the column and the reduction in inertial forces raises the speed of the engine up to a measured 3,000rpm (reciprocations per minute).

There is a large cooling fin just behind the displacer. The engine needs to get very hot to achieve its maximum power and the platform also gets very hot. Because the platform is so long, it sets up its own frequency and the aluminium plate bends and flaps. It is now reinforced with a piece of $^5/8$in steel to stop the platform from vibrating. These are clamped together with a custom-made nut that also acts as another set of cooling fins. The intention is to fit a linear generator and the extra cooling would be a necessity. Its wooden base is painted silver. The design with the cooling fins above the hot end is not ideal and at one point a stainless-steel shield to deflect the heat coming up through the cold end was fitted and a considerable improvement was noticed.

When the engine was first shown at a show in Brighton it caused quite a stir as it was unique at the time. The engine could have been developed further, but I progressed to other projects.

eventually jumping out. With a longer cylinder it would probably work as expected.

The displacer has a $^3/8$in silver-steel displacer rod which is bored out so that the wall is only a few thou thick. This makes it very light. The hole down the middle was maximized as much as possible. Inside the displacer cylinder the displacer is clamped with O-rings top and bottom.

The power piston is attached to the beam by a Bowden cable. On the end of the beam there is a long stainless-steel tube with two brass weights at the ends of the rod. The movable weights on the end of the beam allow the reciprocation frequency to be altered. This is similar to the way in which a figure skater can alter their rate of rotation by altering the position of their arms. The beam was reduced in weight in order to minimize the inertial forces on the piston. The Bowden cable bends in operation and this helps with the timing (a curve can be seen on the cable when a strobe is put on it). When it is running at high revs the bow in the cable allows

6 Low Temperature Differential Hot Air Engines

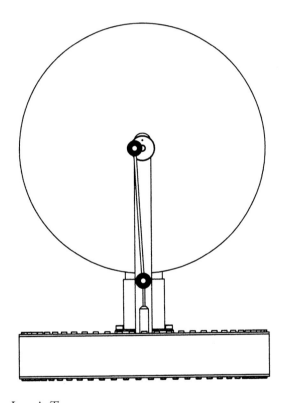

Low Δ T.

Low Δ T (Low Delta Temperature) engines commonly have a displacer cylinder differential swept volume of about one hundred times greater than the power piston, with the equivalent size pistons to suit. The difference between the volume of the power cylinder and the displacer cylinder allows it to work using a very low temperature differential.

Most Low Δ T engines can be instantly recognized by their small power cylinder sitting next to a large displacer cylinder. These engines have large flywheels that are light and often use intricate parts for the connecting rods in order to keep the weight down.

Professor Ivo Kolin did a lot of the early work with Low Δ T engines. One of his notable engines had a square displacer piston and used a diaphragm method of replacing the power piston. Professor Senft PhD, Professor of Mathematics, has done a lot of excellent work on the theory, mathematics and reduction of the temperature differential that these engines require to operate. Justin Jones wrote an interesting article in *Model Engineer* in 1995. At the time, his Low Δ T was the best I had come across and it was the first example I had seen that was able to run from the heat of one's hand. Bob Atkins has since taken these engines a step further by making them with precision-ground graphite pistons that are ground to within microns and used in glass cylinders.

My own contribution came quite soon after I saw a photograph of Professor Senft's engine. It was from this that I produced my Ten Inch version, which also has a cylinder swept volume one hundred times greater than the power cylinder. Visitors to the Brighton model engineering show in 1994 would have seen for the first time in England a Low Δ T engine running. My Ten Inch Low Δ T engine created quite a stir with the general public at the show.

EIGHT INCH LOW Δ T

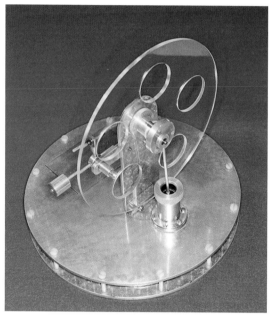

Eight Inch Low Δ T with power piston in the foreground and lightened perspex flywheel.

I made this engine in 1995 to present to a French museum, La Musée de la Blanchardiere at Flers in Normandy. I have previously exhibited at shows in this museum and I made this presentation engine for them on behalf of the Worthing Society of Model Engineers. Low Δ T engines were not very common at the time and the museum was thrilled to receive it. It stood in an oak and glass case in their restaurant for several years until a fire destroyed the museum and everything salvaged was put into storage. When I returned in 2002 the mayor of Flers returned the engine to me and it is now back in my collection. It wasn't damaged by the fire, just in need of a thorough clean due to dust and dirt getting into the bearings.

This Low Δ T differs from others I have seen and made in the way that the displacer mechanism operates. Instead of going up and down gently, the displacer disc has a jerky motion where it is held up against the cold end and then dropped abruptly on to the hot end. The displacer cylinder comprises a perspex ring between two $^{1}/_{8}$in thick, 8in diameter aluminium hot and cold plates that are held together by twelve nylon screws. O-rings are let into the plates and clamp down on to the perspex ring to seal it, making it airtight when the nylon screws are tightened. Other Low Δ T engines often use silicon sealant on the plates to adhere them to the ring, but this method of locating and sealing the ring allows it to be dismantled for cleaning. I 'engine-turned' (known as jewelled in some circles) the aluminium plates to give an attractive finish.

The polystyrene displacer piston has a brass guide with small bronze bushes inserted at either end. The purpose of this is to reduce friction. To have a bearing all the way through the guide would require a very small hole, which would have been difficult to drill straight. Drilling a larger hole through the guide and putting a bush at either end reduces the potential friction.

The power cylinder is made of cast iron with an aluminium piston running in it. The cylinder has an aluminium case and a cast-iron liner held in by a knurled screw at the top. The drive train has four ball races, one to drive the displacer via a balance crank, two in the main bearing housing and one on the big end of the power piston.

The main bearing housing is screwed and locked with a locknut into the main column, making it slightly adjustable. The column is shaped from $^{1}/_{8}$in thick aluminium and secured in position by $^{3}/_{8}$in square section aluminium bar, bolted from the underside to the cold plate of the displacer cylinder. There are two small softened cross-drilled balls that have been drilled through and fitted to the con-rod from the crank pin. These balls act on the lift and drop mechanism that pushes the pivot to raise and

A reverse view of the engine showing the counterbalance weight, the displacer piston con-rod and the presentation plaque.

Ten Inch Low Δ T. At the front the top heat sink is visible. Note the method of fastening the top plate to the yellow plastic displacer cylinder wall.

lower the displacer piston. The balance arm that pivots in plain bearings has a balance weight to offset the weight of the displacer piston. If these bearings were changed for ball races, the performance of the engine would improve.

The perspex flywheel is lightened with five holes drilled around a central PCD (Pitch Centre Diameter). It is mounted centrally between the two bearings to ensure that the weight is equally distributed over the two ball races. To achieve this there is an offset spacer on each of the three screws that attach it to the crank disc.

TEN INCH LOW Δ T

The power cylinder is made of stainless steel and was made from another scrap cylinder. It proved unnecessary to bore it out. The cylinder was lapped and polished to a high finish and a flange was fixed with Loctite on to the bottom. It has six fixing holes and a gasket and is attached to the displacer cylinder with six screws.

Graphite is used for the power piston because it requires no lubrication. The graphite power piston is turned on the lathe until it just enters the cylinder and has a very good surface finish. The proper fit is achieved by working the piston up and down the cylinder until it frees up. It is done this way because graphite cannot be lapped. More detail of this operation can be found in Chapter 16, 'Workshop Practice and Techniques'.

The displacer cylinder is 10in OD (Outside Diameter) yellow plastic gas main pipe. There are two 10in diameter duralumin discs top and bottom. These are held to the plastic with sixty 6BA screws either side. The holes for these screws are drilled so that the top and bottom fixing holes in the plastic are offset by half the distance between the holes. The offset of the top and bottom holes was necessary to avoid through-holes in the plastic.

This shows the finish and fit of a graphite piston.

The plastic displacer cylinder has been shown clear to demonstrate the offset screw method of attachment.

The polystyrene displacer piston was made by temporarily attaching it to a plywood faceplate before placing it in the lathe. It was cut from some packing material and is held on a 1/8in diameter silver-steel rod that goes through the polystyrene. This rod is threaded at the end to enable the attachment of two threaded plates that clamp the polystyrene. This helps to keep the expanded polystyrene displacer piston true and square in the bore.

The aluminium clevis that holds the piston is screwed in from the centre of the power piston with a countersunk screw and held in with Hermatite-type gasket sealant to give a good seal. A 1/8in diameter pin runs as a pivot in the clevis. The aluminium con-rod's little end has a phosphor-bronze bush that connects to the clevis via the clevis pin, and the big end on the same con-rod has a flanged ball race.

The drive mechanism is a crank disc with a crankpin set into it. It has a ball race, held in place with a 6BA nut. It is screwed into the crank disc and is positioned to give the desired stroke. It also operates a lever that just rocks the counterbalanced arm.

The flywheel is a piece of aluminium from an old aeroplane at Shoreham airport. This 1/8in thick duralumin forms a 10in

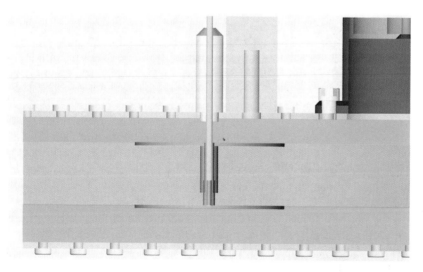

This diagram shows the clamping method used to secure the Styrofoam displacer piston.

The con-rod on this Bob Atkins Low Δ T has a bearing at one end and a bush at the other. Note the method of balancing.

A selection of ex-computer bearings.

cylinder with countersunk screws and sealed with a Hermatite-type gasket sealant in the two tapped holes.

When this engine was first made and stood above boiling water, the two ¹/₁₆in duralumin plates bulged. Additional cooling fins were added on the top to improve performance. It is now powered from the heat of a 40W bulb placed at the bottom of a 5in diameter metal tube. This keeps it running at shows without attention.

diameter disc. It may be of some advantage to add a few holes so as to reduce its overall weight and increase it relatively in the outer ring. The bearings and the 5mm shaft were from an early computer. I have used a lot of these ex-disk drive bearings from early computers because they are very well engineered and already have bearing housings. The crank discs are held on with 4BA grub screws. One is fixed with Loctite on the shaft and the other is adjustable. This is required to allow assembly and to adjust the timing.

The flywheel support is a ¹/₈in plate, attached to a ⁵/₈in square bar. This is attached at the bottom from the inside of the displacer

The cooling fins are attached to the duralumin disc of the displacer cylinder and heat transfer paste has been added.

It is engineered well, but I used techniques in construction that are more appropriate to making an ordinary hot air engine rather than a Low Δ T, such as ball races and large diameter shafts, which make a heavy construction for a Low Δ T. In this engine's early days I designed it to run over a Pyrex bowl containing hot water. This was supported on a tripod. It would run for about three hours on a bowl of hot water before the water cooled sufficiently to a state in which the temperature differential had reduced enough to stop the engine.

One of the biggest difficulties with this engine was getting the balance just right. It took about a day, and it was done in two parts. The weight of the displacer on the end of the arm was balanced as a separate unit.

When balancing the power piston I used a counterbalance. When balancing the other side a moveable weight was clamped on to the crank disc. The weight was variable in rotation and length. This balanced the weight of the displacer piston to the point where it will stop anywhere on its revolution, giving a good indication that the balance is about right.

Bob Atkins counterbalances his engines by simply adding a piece to the flywheel. His engines are some of the best I've seen.

Bob Atkins' method of balancing his Low Δ T.

Bob Atkins' Low Δ T entire engine. An excellent example of a well-engineered simple Low Δ T.

Balancing mechanism clamped to the crank disc.

Large versions about the size of a mattress have been used to pump water. Others float on the water that needs to be pumped. The water provides the cold side and the sun provides the hot side.

7 Martini Hot Air Engines

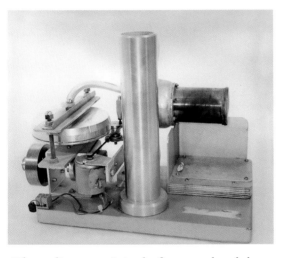

The cooling tower is in the foreground and the hot end on the right. On the top left is the diaphragm chamber and on the lower left is the motor that is used to drive the displacer piston.

This shows the top of the diaphragm chamber, the electrically driven displacer rod and the air transfer tube. Note the reinforcing angle section on the top of the diaphragm chamber to stop it flexing.

Martini engines can be recognized because they have a displacer that is not mechanically coupled to the power cylinder. The displacer is electrically driven and this enables the speed of the displacer to be controlled, from slow running up to several hundred rpm. The speed of the displacer therefore determines the speed of the engine. The Martini engine cannot be started by hand, which may seem like a bad idea, but it is because of this that it is such an interesting engine, with countless possibilities for all sorts of applications.

You should set a slow speed to start with and allow the set frequency to be picked up by the power piston. Once it receives the pulse from the displacer it starts to rock. Eventually the flywheel throws the diaphragm or piston over and the engine starts on its own and in the direction of its choice. The speed can then be varied, but this has to be done smoothly or it will stop. This does not work with any other Stirling engine because there is no way to control the speed other than with direct dynamic heat application.

With another housing for the diaphragm it could be made double-acting, with two

This view shows the eccentrically mounted ball race from the crank and the drive from it to the power diaphragm chamber. (Reproduced with the kind permission of Graham Darlington)

displacers, one on each side, which would be very powerful. A sprung diaphragm could be used, but there would be little, if any, benefit. The power put into the spring to compress it would result in slightly less power back again when it returns. Overall, in this engine, power would be lost. When the pulse is received from the displacer the power diaphragm is deflected and being 1/16in thick rubber it does in a sense spring-return. When the pulse pushes it down you have to overcome the springiness of the diaphragm. Diaphragms are usually made from much thinner material.

The original idea was to make a boat engine, but this never happened and it became purely experimental. The engine is quite powerful but not easy to start. The large diaphragm is clamped by a 1/8in aluminium plate that holds the rubber against the machined aluminium diaphragm chamber. The frame holding the diaphragm chamber flexed during operation so some angle section was added to the top to reinforce it. Without this, some of the motion that should have gone to the shaft would have been spent by bending the frame that was holding the diaphragm chamber. The big end of the crank is an eccentrically mounted ball race.

This shows the underside of the diaphragm chamber and the flywheel.

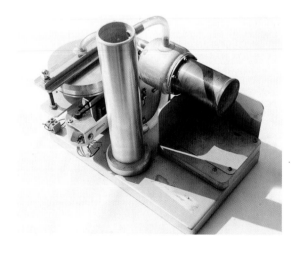

The cooling tower, the hot end, the sugar bowl water jacket and the motor for the displacer drive are shown here.

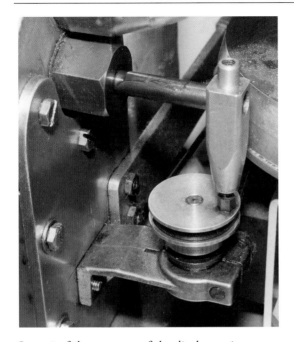

Stage 1 of the sequence of the displacer piston drive from the electric motor.

Stage 3.

Stage 2.

Stage 4.

The drive to the stainless-steel displacer has a rocking, reciprocating motion which agitates the air inside and gives better performance due to more turbulence. I originally saw this type of drive on an engine made by my close friend Geoff Bartlett, who uses it successfully on many of his unusual engines.

A scotch crank could also be used instead of the mechanism discussed here, but I prefer this method as it is quieter.

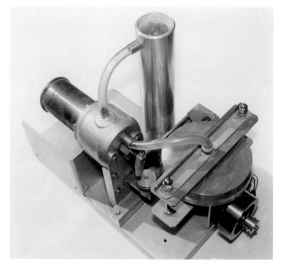

Note the interesting drive for the displacer on this view of the complete engine.

The 12V motor that powers the displacer takes very little current. The power generated is sufficient to run a generator that powers a small electric fan. Far more power was generated than was required to drive the displacer and hence plenty of useful effort is available. Several experiments were done to try to make it power itself electrically. Intricate electronic circuits were put together by my son, Graham, but every time we switched over to automatic operation the frequency was lost and the engine stopped. With more perseverance, I am convinced it could be done.

The main shaft runs in ball races of the type used on early computers with a 5mm shaft. It currently runs up to 600rpm. It is water-cooled by a convection system, with the cooling tower feeding the water jacket from the top. The water jacket then connects to the base of the tower. The stainless-steel water jacket was originally a sugar bowl. This was bored out, glued in position with Araldite, then sprayed silver. The hot cap is made from stainless steel.

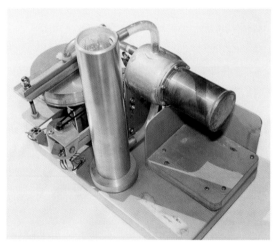

This shows the hot cap and the water jacket assembly.

8 Marble Engines

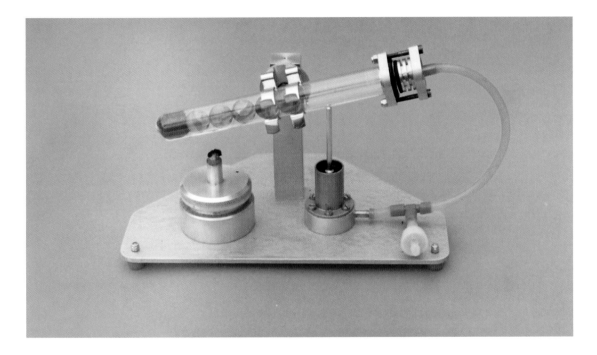

The marble engine is one of the simplest forms of Stirling engine. With these little nodding engines you can observe exactly what is happening, since the displacer cylinder is a Pyrex test tube and the displacer piston is replaced by ball bearings or glass marbles. The marbles move from one end of the tube to the other as it tilts. The test tube is pivoted in the middle and rests on the power piston extension rod.

Marble engines are excellent for demonstrating the Stirling cycle. When the air is in the hot end it expands and pushes the piston rod up. This tips the tube and the balls

Marble Mk II in early configuration.
(Reproduced with the kind permission of Graham Darlington)

roll to the hot end. When the balls are in the hot end, the air goes to the cold end where it contracts, making the piston drop back down and allowing the test tube to tip back again following the piston rod down. The air then returns to the hot end where it expands again and the piston rod will rise, tilting the test tube once again. This is the Stirling cycle and it keeps going for hours on end.

I made my first marble engine in the winter of 1995 and the first one I put on show was at Olympia at Christmas of the same year.

The test tube is gripped in a suitably sized Terry's Tools clip, which enables the tube to be moved to adjust to the optimum running position. The Terry's clip is screwed to a short shaft, which in turn is supported by two small ball races in a housing. It is a fascinating engine to watch as it rocks backwards and forwards.

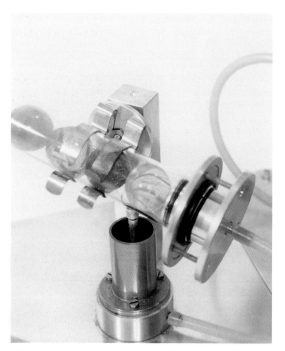

This shows how the test tube (displacer cylinder) is secured to the pivot with a Terry's clip. The optimum balance can be achieved by moving the tube through the support clips.

These engines are powered by a small spirit flame that is applied directly on to the glass. In experiments, various fins have been tried, but glass is such a good insulator that they are little more than a counterbalance weight. After running for a while, the temperature does actually rise and it has to be rebalanced. (For

this reason I designed my later marble engine with a flywheel, which eliminated the need for this adjustment.) The position of the test tube can be adjusted by watching the movement of the marbles. If they seem to be hitting the steel wool at the hot end more sharply than usual, the tube should be moved towards the cold end. If they seem to be hitting the cold end rather more heavily, the tube should be moved a little towards the hot end. Once this has been mastered the little engine will keep running until all the fuel has gone.

MARBLE MK I

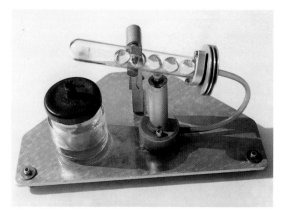

Marble Mk I. (Reproduced with the kind permission of Graham Darlington)

This is the first marble engine I made. Attached to the small aluminium base plate is the power cylinder with a piston attached to a solid rod. The tube is held by a Terry's clip, which in turn is set into ball races so that it is able to rock a few degrees either way. It is pushed up by the power piston con-rod. As the engine has very little power it is desirable, although not essential, to use graphite for the power piston as it works with little friction and doesn't require oiling. However, I have made pistons of cast iron, brass and aluminium and these have all worked, so it isn't like the Low Δ T where graphite is essential. Cast iron is ideal for a cylinder even when using the graphite piston.

Using three feet on these engines means that if the surface an engine is stood on is not quite level or is uneven the engine will remain stable, like a three-legged stool.

The con-rod is attached to the power piston (there is no need for any flexible joint) and it is screwed directly into the power piston. I glued a ball bearing with Araldite in the top end of the con-rod so that it glides smoothly on the glass. The con-rod is hexagonal in section and requires a guide. There is a round hole at the top of the guide where the edges of the hexagon run, causing less friction as there are only six points of contact with the edge of the hole. As well as being a guide for the con-rod the hole also allows the air to escape as the piston comes up. If a round hole and a round rod were used a hole would have to be drilled to allow the air to escape. There is no little end as the con-rod is attached directly to the power piston.

The only bearings are the main bearings, which are two small ball races through which a spindle runs. This, in turn, holds the Terry's clip, allowing the test tube to rock. I advise ball races, but if you haven't got any you could use a small shaft and brass bushes. Plain bearings such as Oilite bushes could be used, although it is far better to let it rock on ball races. The model discussed here is an early marble engine so there is no flywheel.

The displacer cylinder (test tube) is coupled to the power piston and cylinder by a small flexible tube. Everything is attached to a triangular $1/8$in aluminium base plate. There are three cooling fins on the back which add extra weight at the cold end, keeping it in contact with the power piston.

At one end there is a flame and this becomes the hot end. The engine is fired by a methylated spirit burner that only produces a small flame on the hot end.

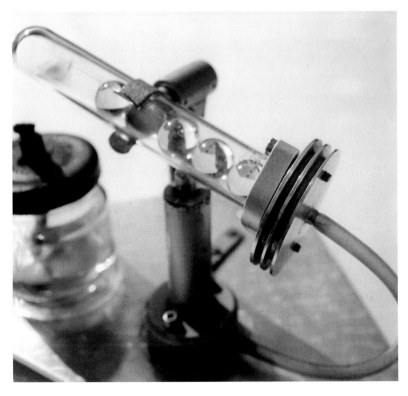

A simple method of mounting cooling fins to a test tube.

> Don't let surplus oil get back into the test tube. I have used several methods of sealing the opening and have found that when using the larger tubes with aluminium bungs fitted with O-rings and even with plain rubber bungs they can be blown out. I have ended up having to deal with several very hot balls rolling around on the table top and falling on to a vinyl floor, leaving indelible tracks. Even worse is the automatic reaction to catch them when you see the bung coming out.

MARBLE MK II

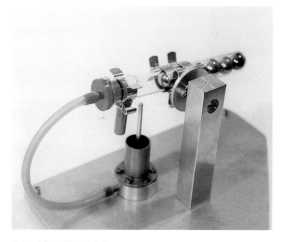

Marble Mk II. The entire engine.

This is the second marble engine I made. It has ball bearings instead of glass marbles, which roll better because they are precision-ground. It has been demonstrated at shows all around the country. The performance is similar to the Marble Mk I engine, except that the balls roll more freely. It has an open-ended cast-iron power cylinder held on to an aluminium block with 6BA screws and has a highly polished cast-iron piston.

The base is a rectangular $1/8$in aluminium plate on three feet. It has a heavy $3/4$in square pillar, drilled and counterbored to accommodate two ball races. There is a disc attached to the shaft and there are two Terry's clips

screwed to the disc. At the rear there is a plastic plug with an O-ring in it which provides a good seal.

The light aluminium con-rod is glued with Araldite into the power piston and is domed at the end to push against the glass. It has a brass counterweight, which was added as the ball bearings were heavier than the original marbles. It also helps to keep the glass in contact with the con-rod.

There are no cooling fins. Being glass the heat doesn't travel much so the cold end doesn't get very hot.

MARBLE MK III

Marble Mk III is a larger version of the marble engine. The cylinder is brass with a cast-iron piston. The cylinder is attached to the aluminium base and cross-drilled to allow the air in at the bottom. It is honed and lapped.

The outside diameter of the tube is 1in. Currently it has marbles in it but it has used ball bearings. The advantage with ball bearings is that it is possible to get any size you like; the disadvantage with glass marbles is that they tend to chip and break up with the heat. At the end of the test tube is ordinary wire wool, which acts as a shock absorber and retains heat at the hot end, a bit like using a regenerator.

The con-rod is aluminium, glued with Araldite into the base of the cast-iron piston. There is a ball bearing set into the top of the con-rod; this enables it to slide on the glass better. The connection is a small piece of flexible plastic tube to carry the air from the displacer tube down to a hole at the bottom of the power cylinder. This hole is drilled horizontally in the base and another is drilled vertically into the middle of the power cylinder.

A $1/8$in triangular aluminium base with three feet holds the main $3/4$in square aluminium column that contains two ball races. A crank disc is attached to the column with a shaft running in the ball races. This one has two Terry's clips, which hold the tube more securely than a single one.

Note how the test tube rests on the power piston.

Side view of the engine.

Marble Mk III. This shows the wire wool at the end of the test tube.

The engine has an aluminium plug with an O-ring in it and three cooling fins, one of which goes over the neck of the test tube. The first fin is a clearance hole over the glass tube with an O-ring between the glass and the aluminium. The second fin holds the bung in to stop it blowing out. The performance is not as good as with the smaller engines and it is temperamental in running. The heating is provided with a small methylated spirit burner.

This shows how the O-rings are used to seal the engine.

Left *This shows the pivot disc and the bearing mounting.*

The sizes of the marbles available often are not a suitable diameter in relation to the inside diameter of the test tube. I went to a marble shop where they had bins full of marbles. I tried one in the tube for size and it got stuck, so no sooner had I got there than I was in their staff canteen putting hot water on the tube to try to blow the marble out. I told them what I wanted and even though I only intended to buy half a dozen they got lots of bags out and there were marbles all over the floor. It was just before Christmas and there were people all around stepping over the marbles – what a wonderful hobby.

FLYWHEEL MARBLE

This engine was designed and built specifically for demonstration purposes. It has a flywheel so as to eliminate the need for regular adjustment when running all day at shows. Adjustment would be necessary to compensate for the reduced efficiency of the engine as the cold end heats up. The flywheel stores the energy to keep the momentum going through the cycle.

The Pyrex displacer cylinder has five stainless-steel ball bearings. It was originally made with marbles and a wire wool buffer at the end of the test tube, but ball bearings retain the heat and removing the wire wool has stopped the ball bearings becoming fouled with bits from the wire wool.

The graphite power piston runs in a bronze cylinder. The scotch crank mechanism serves to pivot the whole of the displacer cylinder (test tube). The power piston con-rod operates on the crankpin as the crank disc rotates. The crankshaft and crankpin are made from silver steel and the flywheel was machined from $1/2$in thick aluminium with a rim left at the full thickness. The centre was machined down to reduce the weight. In order to maximize the benefit of the energy stored in the flywheel, its speed was increased by using a gear ratio of 2:1, which means that the flywheel spins at twice the speed of the crank disc. The large gear is attached to the crank disc and the smaller gear is attached to the flywheel bearing assembly.

The black-painted aluminium base is 12in long and $5^{1}/2$in wide with four rubber feet

Flywheel Marble.

screwed into it. Attached to the base by means of a square section block is a frame which is cut from the same material and has fixings for the crank disc, the flywheel bearing and the pivot bearing, which in turn is attached to the Terry's clips. The Terry's clips hold the test tube.

LONDON MARBLE

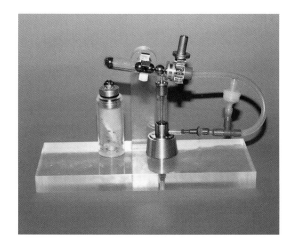

London Marble. (Reproduced with the kind permission of Graham Darlington)

This engine was made by a friend who donated it to me and it is now part of my collection. The power piston can be seen moving up and down through the glass cylinder. This glass cylinder has a leather washer piston of the type found in cycle pumps.

There is a rubber bung in the end of the tube and a bleed valve that was attached later. When running marble engines at shows it helps to fit a non-return valve, the sort that can be purchased at an aquarium shop. If there is any leakage of air it takes in fresh air and on the suction stroke it supercharges the system. The test tube is only 3in long and $^1/_2$in diameter. It carries four small steel ball bearings. There is a single Terry's clip with two ball races let into the perspex column.

On top of the glass cylinder there is a brass cap with three small holes to allow the air to

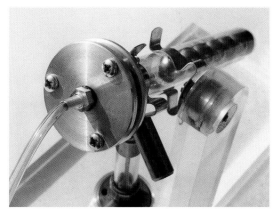

This shows the bearing pivot assembly in the Perspex support column.

escape. The central hole guides the con-rod, which is a piece of plain silver-steel rod of $^1/_{16}$in diameter.

The base and column are perspex, giving it a completely different look to my marble engines. It is also very much smaller. The glass power cylinder is held in an aluminium base, which in turn is attached to a perspex base. There is a small aluminium cylinder attached to the cold end with four small cooling fins. It is powered with a small methylated spirit burner.

> My retirement has been taken up with demonstrating hot air engines and explaining the principles of how they work, often to be greeted by looks of disbelief. One lady once said to me 'Even if what you say is true and I can see it working, I still don't believe it.'

AMERICAN MARBLE

At the Brunel show I met an American who had read an article in the *Model Engineer* about my Marble Mk I engine. The title of the article was 'Noddy. A Hot Air Engine for Everybody'. He thought it was a fascinating subject for mass production, so instead of making one he decided to make dozens. On a laptop he

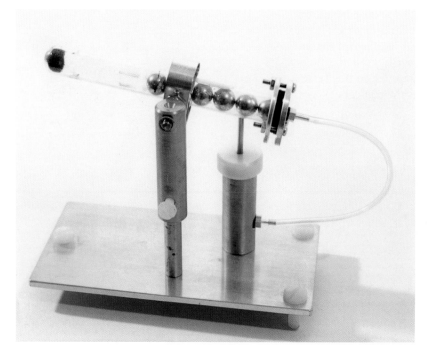

American Marble. Note the nylon bearing guide and cap on the power cylinder. The adjustable nylon levelling feet are simply screwed into three threaded holes in the base.

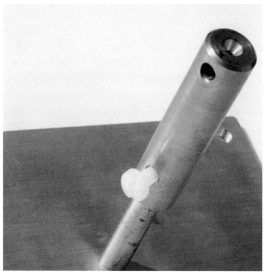

The adjustable column and its locking screw.

parts on a bench awaiting assembly and views of all his lathes set up to mass produce cylinders and pistons. I was very interested and honoured that he had chosen my 'Noddy' engine to mass produce. He told me later that when he had finished the engines he copied the article in the *Model Engineer*, added some of his own material and then presented an engine along with the article to all of his friends and relatives. At a later date I received a box from America and in it I found that he had presented me with a 'Noddy'. He even sent a spare tube and a spare length of wick.

The brass cylinder is attached directly to the aluminium base. The leather piston is lightly oiled and the con-rod is rounded where it pushes and slides against the glass of the test tube. The guide at the top of the cylinder is a nylon plug that sits over the outside of the cylinder. The con-rod comes up through the nylon and there is a large clearance hole to allow the air to escape. The base is 1/8in aluminium plate with a bevelled finish.

The three screws and alloy ring have a clearance fit over the tube and an O-ring on

showed me his version of the 'Noddy'. It had a rectangular base on three nylon feet with nylon screws screwed from and through the top so that the base could be adjusted. On the laptop were pictures of rows of all the component

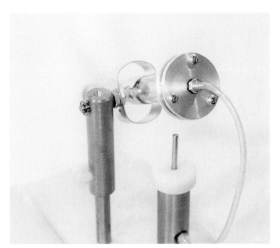

Close up view of the end of the air tube that connects to the displacer cylinder.

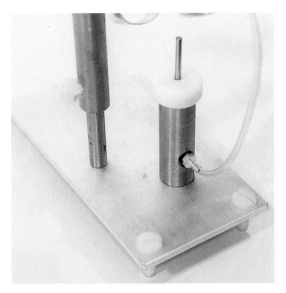

Close up view of the end of the air tube that connects to the power piston.

the neck of the tube. The other plate holds a tube which runs to the cylinder. The bearings are a plain shaft through the solid brass column. There is a tool clip which must be of American origin fitted to the shaft which holds an ordinary Pyrex tube. There are two aluminium fins at the back, holding the plug in, similar to my description in the 'Noddy' article.

The top of the brass turned column is adjustable on the bottom part and there is a knurled screw set into it so that it can be moved up or down and then locked on the bottom post.

9 Unique Twin-Cylinder Boat Engine

This was my first twin-cylinder boat engine made in 1996. It is a unique design. The concept was to design a twin-cylinder engine with only one linkage, but initially I did not know how to get the phase right. I spoke to Ken Boak, who is heavily involved with the Stirling Engine Society. He suggested crossing the air over from one displacer cylinder to the opposite power cylinder and to put the air behind the piston instead of in front of it. It took another year to produce the engine due to the complexity of the internal design. It required a lot of skill to fit everything into the available space and also to make it a free-running engine.

There are four pistons operated on two rods. On each rod is a power piston and a displacer piston. The air from the displacer piston crosses over and goes to the base of the opposite power piston. It is necessary to have a power piston on a rod and a central bearing with the displacer on the end. Mechanically, this is a difficult thing to achieve.

The linkage is a Ross linkage modified to my own design. There is no swing arm but something similar to a scotch crank that the centre bearing runs in. A ball race runs in-between two plates. It looks erratic when running, but the two rods are in parallel motion. Everything has to be concentric and the

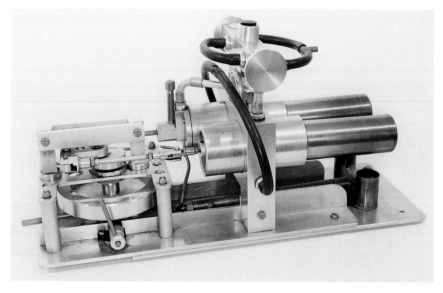

Unique twin-cylinder boat engine.

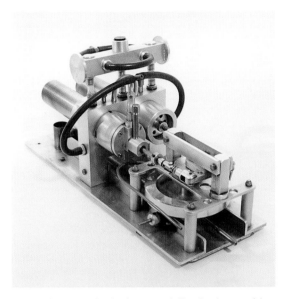

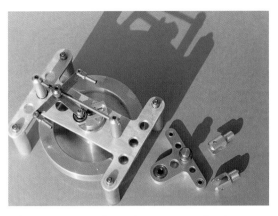

The components shown here go to make the modified Ross linkage assembly. (Reproduced with the kind permission of Graham Darlington)

View showing the linkage and flywheel assembly. Note the difference at the rear of the two cylinders.

rods have to be perfectly true. Silver-steel rods are often not to spec. They can be warped or bowed. It is worth buying silver steel of a well-known type such as Stubbs, which is guaranteed to be the specified diameter and straight.

The 1in wide central block has seven ports that are used for air, water and oil. The oil ports lubricate the guide bushes for the displacer. The air ports transfer the air between the cylinders to make this, unusually, an engine with the power and displacer piston on the same shaft. The water ports are connected to a plenum chamber, which distributes the water for cooling.

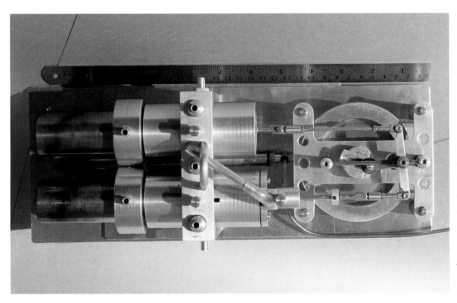

Plan view of the engine showing the Ross linkage and con-rod layout. (Reproduced with the kind permission of Graham Darlington)

Main block and cylinders. (Reproduced with the kind permission of Graham Darlington)

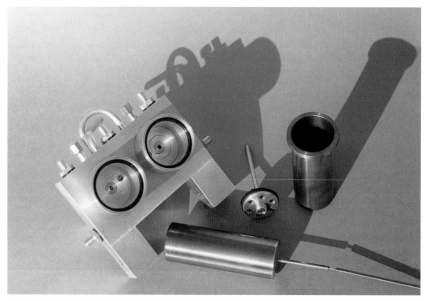

Here is a view of the displacer, power piston and cylinder, giving a clear indication of the layout of the various ports on the block. (Reproduced with the kind permission of Graham Darlington)

The flywheel is horizontal, another unusual part of the design. Underneath the flywheel is a friction drive. If it was put in a boat, the friction drive underneath would be radio-controlled to give forward, reverse and neutral. The shaft comes out in line with the engine and would run straight to the prop shaft with a flexible drive.

The engine starts very easily. Once the flywheel has been turned until it reaches TDC (Top Dead Centre) of one of the cylinders it snatches the flywheel out of your hand, almost self-starting.

The burner is a single mixing tube that goes to a T-piece at the burner end, where the air and gas mixture splits left and right to separate

A view of all major assemblies and parts. (Reproduced with the kind permission of Graham Darlington)

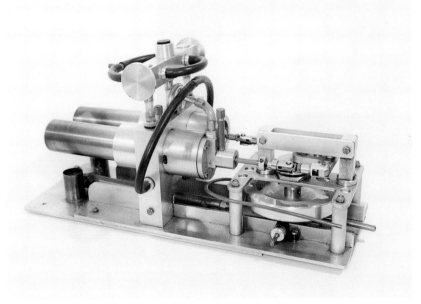

The burners on the left can be seen under the displacer cylinders and the flywheel is on the right.

burners. A better design would be to have two mixing tubes with a T-piece at the gas end so that gas is supplied to two jets. Each burner would get its own gas and air supply and the mixing would be more effective.

If the engine was in a boat it would be tipped up slightly at the front to give the correct angle for the propeller. At the time of its construction this engine was a big achievement, being my most complex engine so far, although the very complexity of the engine deterred me from putting it in a boat. There is no reason to consider it unreliable except that the more complex an engine gets the greater the chance of something going wrong.

10 Stirling Silver I (Twin-Cylinder Gamma Engine)

With my grandson Laurance proudly holding Stirling Silver I. (Reproduced with the kind permission of Graham Darlington)

Stirling Silver I is 50in in length with a 12½in beam and is 6in in depth. It has a twin-cylinder Gamma engine with a modified Ross linkage, twin burners, twin flues and twin water jackets. The cooling water is contained in an onboard tank and is pumped around with an electric pump. The hull is far bigger than it needs to be and the first 4–5in at the front of the boat are solid timber. It took a whole day to carve the square block into shape. If it was half the weight the performance would improve, but even so it looks quite majestic going along. It is powered by a butane cylinder. It appeared on the Internet in several places and was a forerunner of Stirling Silver II, which is a far

superior boat due to its more streamlined shape and powerful engine.

There are water jackets around each cylinder at the cold end and the main block also houses the power cylinders. The aluminium plates – two of them ¼in thick, one ¾in thick and one ³/8in thick – have paper gasket material between them. This lets the maximum amount of water around the cylinders, thus reducing the temperature of their cold ends effectively. The main power cylinders are glued with Loctite into the block, which makes it an easy construction and enables the water to flow around them. This would not be possible if it were a solid block. The two flues have a row of

Stirling Silver I under way in the water.

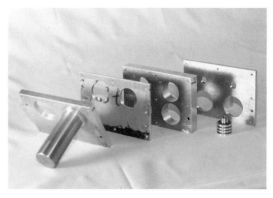

Components that go to make up the main block.

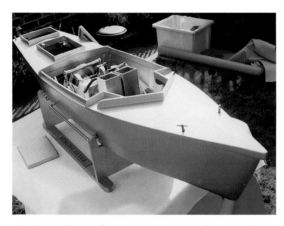

Stirling Silver I front view. A view showing the complete boat.

Main block assembly.

air holes in the bottom to give a good draft. There is a square stainless-steel restriction bent to a semicircle to make the hot gas and flame follow the shape of the displacer cylinder, with a cut-out in the top to allow the exhaust gases to escape. The hot gases from the burners are directed around the 0.020in thick hot end, which is stainless steel. This engine uses leather washer pistons similar to the type used in cycle pumps. They have run for years with very little lubrication. This type of leather piston must never be allowed to dry out.

View of the engine as it sits in the centre of the boat before the hull was completed. (Reproduced with the kind permission of Graham Darlington)

Position of the modified Ross linkage stage 1.

Position of the modified Ross linkage stage 2.

There are two ball races in the T-piece of the Ross linkage. The tops of the linkage are bronze bushes and instead of having a swing link, which is the normal procedure, there is a straight rod attached to the main frame of the engine. This guides the T-shaped part of the modified Ross linkage and keeps parallel motion. The top part of the linkage drives the displacer and the bottom part of the linkage drives the power piston rods that are connected directly into the leather piston in the highly polished aluminium cylinders.

There are five ball races in the main shaft, one in the middle and two at either side. The crankshaft is made up of two separate parts, which enable the crankpin to run in the crank

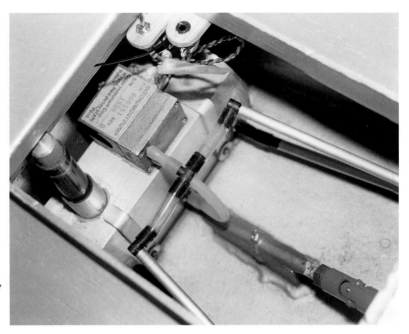

In this hatch are the water pump, the transfer pipes and the water tank. The two universal couplings take the drive from the engine to the propeller.

disc to take the load. It was necessary to put a $1/4$in aluminium plate between the two cylinders in order to take the thrust.

It has a centre bearing with two discs to share the load. It is quite a big engine with twin aluminium flywheels, $4^1/2$in diameter and $5/8$in wide, machined out to make them lighter. The whole engine is quite light.

When the engine was first made the side frames were made of $1/8$in aluminium plate and the thrust of the engine twisted the frame. The engine had to be almost completely rebuilt with aluminium angle, with a plate in the bottom and a tie-bar across the two side plates, plus a central column.

The water tank and pump are all in one hatch. It pumps the water out through two $5/16$in brass cooling tubes. The tubes go through the side of the hull, through the water and back through the hull into the tank again. The tubes either side of the boat prove to be enough to cool the engine sufficiently.

This cooling system is very similar to a car, with the pipe in the water acting as a radiator. The electric pump is a small commercial pump, but although it is small it makes quite a lot of noise. The block has a temperature gauge, which although not really needed is interesting to see what is going on. The main Stirling engine makes no noise at all and cannot be heard when the boat is on the water, but the pump can still be heard several hundred yards away. It copes well against choppy water and weeds are no problem.

Only distilled water is used in the cooling system. The design could be improved by using a different cooling system that would not carry

View of the boat showing one of the cooling pipes.

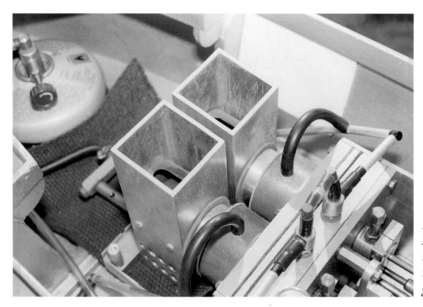

In the chimneys you can see the flame guides. Note the protective heat shield matting in front of the gas cylinder.

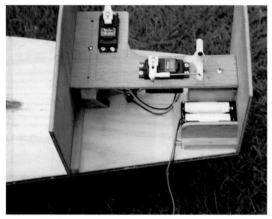

The rear hatch with the batteries and steering servos. One servo is for steering and one engages the drive.

its own water or have an electric pump. Stirling Silver II picks up its own water from the lake, which is a far better method.

There are two 2in aluminium square tube chimneys with a wall thickness of $^1/8$in. The burner is my usual one, with stainless-steel wool and a round pipe-like burner case. It is fed from a single fuel tank that supplies both burners.

The drive from the engine is two flexible couplings. A better design would have been to put the prop shaft at less of a sharp angle. The propeller is not at the correct angle for the water, but not being a boat builder one realizes these things after construction. The radio-control receiver for the boat is located in the hatch, which accesses the water tank and pump. There is a little LED light to show when the pump is on. Likewise with the radio, there is a socket in the main hatch where the batteries can be charged in situ. The rear hatch houses the batteries, switches and radio-control gear.

The drive is the same type as that used by my good friend, Peter Beier from Austria. He has made several boats with a friction disc drive on to the flywheel to get forward, reverse and neutral. This is a wonderful drive. It never slips and is quiet and any future boats will have a similar drive. There is no need to have speed reduction. The engine revs at full speed and when the plates engage the boat gently pulls away. The boat can stop in half its length by just putting it straight into reverse.

This is a big boat with a small displacement. The 6in depth could easily be reduced by half, as it sits on the water with a draught of only

The rear view of the friction disc.

This shows the front view of the reversible friction disc, which gives forward, reverse and neutral.

1¹/₂in. The propeller is handmade, with four very large blades and is 3in diameter. I had the propeller at the start so the boat was made to go with it. It has been suggested that the performance of the boat could be increased

considerably if it was changed to a two-bladed propeller. It would take a very large steam engine to drive such a propeller, but the Stirling engine seems to like the large-bladed propeller and copes with it. However, in an experiment with a five-bladed propeller the performance was impaired considerably. The size of the rudder had to be increased too because it was insufficient for the size of the boat.

> In a lake with lots of weeds in, it goes through like a lawnmower without getting jammed, leaving a trail of vegetation behind it!

This shows the final drive bearing mounting, universal coupling and propshaft.

The weed-cutting propeller.

A view of the bulkheads with the engine mounted amidships.

To make the hull, everything was screwed on to a flat piece of plywood and then the shape of the hull was drawn around it. Everything was screwed on in its position and the hull was built around it. It has a false bottom with two skins to give it shape. A flat-bottomed boat, although not so shapely, would be better for cooling, as it would adapt itself more readily for using the cold side of the Stirling engine. Stirling Silver III will include this feature.

11 Solar Engine (Single-Cylinder Beta Engine)

A close-up view of the complete Solar Engine without the reflecting dish.

A solar-powered engine is any engine that is powered from the sun. The first solar-powered Stirling engine was made in the 1800s. Any form of Stirling engine can be used because all Stirling engines work using heat as their primary source of energy. In this particular example, it concentrates energy from the sun using a parabolic reflector, which is then focused on to the hot end of the Stirling engine.

The complete Solar Engine without the dish.

The air temperature inside the hot end was recorded at 700°C before it melted and ruined the temperature probe which was housed in a 6mm diameter aluminium tube. I was unable to extract the probe as the aluminium tube had melted all around it.

The 18in parabolic reflector has a Beta Stirling engine, of the same type used in Stirling Silver II, mounted behind the dish. The hot end goes through a central hole in the parabolic reflector. It has two sets of cooling fins, one inside the reflector and one at the back of the reflector. These are used for cooling and to hold the parabolic dish in place. The nut that screws on to the cylinder to secure the cooling fins and the parabolic reflector also forms part of the Pyrex dome assembly. This assembly, covering the hot end, is effectively putting it in a greenhouse, which increases the efficiency by at least 10 per cent.

The Pyrex glass dome gets extremely hot and it is difficult even to look at the hot end due to the intensity of the light. The Pyrex bottle, which I cut the neck off to make a glass dome, is ³/₁₆in thick. The Pyrex bowl has an aluminium ring, which is internally threaded. On the outside are two O-rings that are pushed into the glass dome to seal it. It grips so tightly that the glass can be turned and used as a handle to screw the whole assembly together. However, the build-up of pressure could shatter the glass, so three slots were cut across thread, allowing the air to escape. For safety reasons I would advise anyone using a glass dome to include a method of relieving the pressure. Remarkably, when the sun shines on the dish it remains cool and the dish itself also acts as a cooler.

I refer to the glass dome as my greenhouse effect.

The dish itself was manufactured in the USA, but can be obtained in the UK. It is 18in diameter, but it is also possible to get them 24in in diameter, which would be far superior.

This shows the sun focused on the hot end, which is housed in a Pyrex dome.

The cooling fins and cover glass.

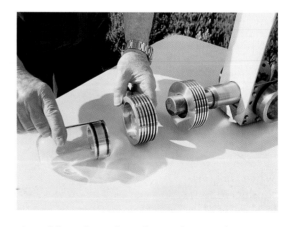

Assembling the cooling fins and cover glass.

*This gives an indication
of just how hot it can get.*

The Solar Engine shown from the rear.

*The aluminium insert in the Pyrex cover and the
pressure-relieving slots.*

When designing this kind of engine the main
consideration is to get the hot end at exactly the
focal point. As an alternative to using a formula
to work out where the hot end should be, turn
a piece of wood so that it has a shoulder that fits

the aperture in the dish. Mount this where the
displacer cylinder assembly will go, making sure
it is the same diameter and is longer than the
displacer cylinder. Temporarily mount the

SOLAR ENGINE (SINGLE-CYLINDER BETA ENGINE)

This shows a close-up of the displacer hot end, the Pyrex cover and the cooling fins centred in the dish. You can also see where the reflected light falls on the displacer cylinder.

wood in place of the displacer cylinder and point the dish towards the sun. The wood will char and smoke at the correct focal point in a few seconds. Measure from the shoulder to the centre of the charred ring and position the displacer about 1/2in further than this, as all the energy needs to be concentrated within the first inch of the hot end.

The complete engine mounted on its tripod ready for operation.

When I first assembled the engine and started it, the only way to focus the dish was to look at the hot end and turn the dish until the ring of light encompassed the hot end. After doing this several times all I could see was an orange glow and I realized that the light was so intense that I could be damaging my eyesight. The idea of using a probe came after this. The probe has a white background which is above the top of the dish and by turning the engine so that the probe turns into the sun it is possible to find the point where the probe doesn't create a shadow. It is then lined up. This is simple and effective. As the sun is 92,750,000 miles away from the earth there is no concern that the centre line of the engine was 10in from the centre line of the probe. Although the angle is not perfectly precise, it is as good as it needs to be for this application. The engine has also been run using two lamps in a photographic studio.

My son Graham observed that if you point a pencil at the sun and there is no shadow then you know that the pencil is lined up with the sun.

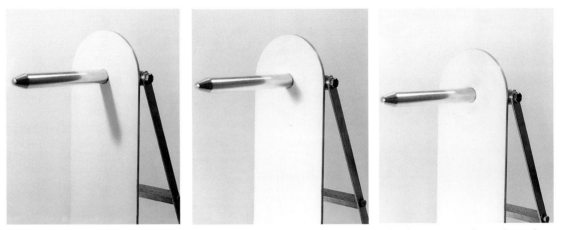

This series of three photos demonstrates that when the shadow of the probe disappears it has achieved optimum alignment.

Above *A close-up view of the engine from the side without the dish.*

Right *Demonstrating the Solar Engine at a model engineering exhibition. (Reproduced with the kind permission of Mike Chrisp)*

12 Stirling Silver II (Single-Cylinder Beta Engine)

Stirling Silver II in the water. (Reproduced with the kind permission of Graham Darlington)

Stirling Silver II is a single-cylinder Beta engine in a hull that was made by Richard White, a marine engineer who lives in Norway. In trials it has proved to be faster and have a longer duration than steam boats, as these regularly need to have their water levels replenished after a short time. Stirling Silver II runs for at least three hours on one fuel canister.

The engine is powerful and efficient and three main factors increase the performance of the boat. Firstly, the performance is greatly improved by the design of the hull. The boat cuts through the water very cleanly and the hull is light. Secondly, the extreme heat differential between the cold side and the heat of the burner returns plenty of power. Even after running for several hours the cold side remains

at the same temperature as the water that is being pumped through the water jacket. Finally, although nothing is touching, the engine runs with a knocking sound. This is caused by the turbulence generated by the closeness of the power piston and displacer, which in turn improves the engine's performance.

The engine runs under full power all of the time. It only has two speeds, full power forwards or full power backwards. When forward or reverse is selected the boat leaps into action. The only way to stop the boat is to put it into reverse, whereupon it stops in half its length.

There is no regeneration, just a straight displacer. The twin flywheels drive on to a

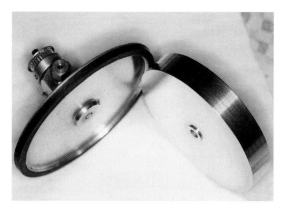

Stirling Silver II friction disc and one of the flywheels demonstrating how the flywheel transmits the drive to the friction disc.

This disc and cap are used to secure the butane/propane cylinder.

friction disc, which is operated by a servo. This gives a very quick response. It has forward, reverse and neutral. A toothed gear on the friction drive drives another toothed gear on the propeller shaft. There is no need for a flexible coupling because the drive belt takes up the difference between the two angles of the friction drive shaft and the propeller shaft. This makes an ideal method for propelling a boat. The small angle of the prop shaft is in this case very shallow and it is not impaired at all by the difference of the two shafts not being perfectly lined up.

The gas cylinder sits on a small ring of plywood which locates it in the stern of the boat and is just forward of the hatch where the electrics are housed. There is a stainless-steel wire yoke that sits over the top of the gas tank and there are two hooks rising from the internal hull of the boat. The fuel tank is held in by two O-rings attached to the hooks and yoke. The yoke is pulled down over the tank to keep it in place. There is a homemade pressure regulator on top of the gas tank that supplies gas to the burner, which is made from 1in square-section stainless-steel mesh. It has an aluminium flue. This engine always starts immediately.

With the gas cylinder being mounted at the rear of the boat it floats with the bow raised slightly out of the water. The performance could be improved with a better weight

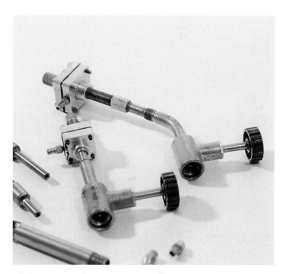

A home-made pressure regulator.

distribution. If the gas cylinder was moved forward by 1in it would alter the centre of gravity, lowering the bow. The boat would cut the water even cleaner with the stern higher out of the water. As it is, with the stern sitting deeper in the water a small amount of extra drag is created, but even allowing for this the performance is remarkable.

The cooling system is used in lots of model powerboats. The engine is water-cooled and

Side view of Stirling Silver II out of the water.

The drive, steering servos, receiver and the water pick-up transfer pipe are seen here.

This shows the power cylinder, the water jacket cylinder and one of the O-rings that seal the water in when they are assembled.

benefits from its own water jacket. The water supply is picked up from the rear of the boat from a small duct very close to the propeller.

The hull is 44¹/₂in long with an 8in beam at the widest point. The depth of the hull is 4in at the bow, 3in in the middle and 2³/₄in at the stern. It is made from four sheets of 1mm plywood cleverly shaped and with only two bulkheads in its construction. These two bulkheads are 5in apart, which allows space for a compartment that houses the battery, rudder, receiver and servos.

The plywood was cut to shape and the rear bulkhead glued on to the two sides. With the second bulkhead spaced at the required distance, the pieces of wood were pulled around and tacked together. Then the bottom was dropped in to give it its shape. The seams were then glued, filled and sanded. You need to be exceptionally dexterous to produce such a hull and it is not a job to be tackled by the average model engineer.

Over the battery compartment there is a thin aluminium lid which clips and slides on to stop water from entering. The front also has a thin aluminium top that has been curved and shaped, glued with Araldite into position and then trimmed. It is sprayed silver with many coats of paint, giving the appearance of a monocoque fibre-glass hull. The aerial runs along the length of the boat in a plastic tube and is fixed with Araldite inside the hull.

It has a 2¹/₂in propeller that revolves at twice engine speed. It is normal to gear down Stirling engines, but in this case the power that is produced means it is not necessary.

Stirling Silver II has spent over 100 hours being tested on lakes and no maintenance has been required except the occasional drop of oil.

13 Buggy (Single-Cylinder Gamma Engine)

This Stirling-powered Buggy, which I also call my 'Moon Buggy' or 'Stirling Crawler', started as a challenge from the Kew Bridge Museum. They held a competition for buggies, with the suggestion of putting wheels on a hot air engine and making it drive itself. It could either be radio-controlled or just a plain buggy to go down a track that they had built for the purpose.

At the time I was in the process of assembling a small experimental air-cooled beam engine. This engine was part-completed so I built a chassis around it and searched for some suitable wheels with rubber tyres. As the buggy was built around this experimental engine, it was a matter of determining the positional configuration of the servos, battery, receiver and gas tank and then designing the width and length to suit.

By the date of the competition the buggy had not been fully refined due to time constraints, but it was still entered into the competition. Subsequently, improvements were made to the burner assembly to prevent the flame-out conditions.

With only a 1in bore cylinder it is not as powerful as it could be and the initial weight of the completed buggy was too great. The weight was reduced by 3lb by drilling and milling areas that didn't need to carry so much weight. This doesn't make the buggy go any faster but makes

The Stirling-powered buggy.

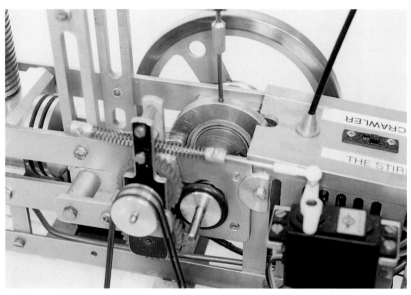

*The ball race
eccentrically mounted.*

more use of the power available. The buggy is capable of climbing up a wooden board with a rise of 4in over 4ft, so it will climb up a board from the road to the pavement over a curb. It is pleasing to watch and the milling actually makes it look more impressive than if it had plain aluminium sides. It is all bolted together, so can be dismantled easily if necessary. It is very manoeuvrable and can stop, reverse and swing around in a small space.

It has a stainless-steel hot cap and displacer. A shroud/baffle over the top of the hot cap guides the hot gases around the hot end and they exhaust up through the chimney. The power piston is completely open at the top, which makes it easy to lubricate. The rest of the lubrication points are also designed to be easily accessible.

The original purpose for building the engine before deciding to convert it to a buggy was to experiment with a drive from the power piston via a beam to an eccentric ball race that forms an offset crank. The eccentrically mounted ball race fulfils the same function as a crankshaft. The benefit of this configuration is that a multi-cylinder engine can be made with a single straight shaft all the way through. As many pistons as required can be driven from this shaft

via their con-rods by using the eccentrically mounted large ball race. The power cylinder is 90 degrees to the displacer and it drives the crank through the beam. For a four-cylinder engine you would have four eccentrically mounted ball races, all with a phase angle difference of 90 degrees. This is a very straightforward method of making a crankshaft and having a single shaft right through means it all runs true and it eliminates any positional discrepancies in the journals.

The link from the beam to the power piston is stainless-steel welding rod approximately $3/32$in diameter. It is not much thicker than a bicycle wheel spoke. The drive from the other end of the beam to the ball race is made of the same material. This is quite sufficient for such a small loading. Forward and reverse drives are controlled by a radio-controlled servo. The drive is kept engaged by means of a spring that maintains a constant pressure. It is geared down and is either engaged or not engaged. This friction drive works well and never slips.

I've never had any problem with this type of friction drive slipping.

Side view of the buggy.

The cast-iron flywheel has been extensively machined away to leave an outer rim. This is the ideal construction for a flywheel as it ensures that all the weight is in the rim, which minimizes the loading on the bearings and performs the best.

The chassis was built from aluminium angle and aluminium sheets $^{1}/_{16}$in thick that are bolted at the front and rear to make a box. The box is required to house the batteries, servos, gas tank and a radio receiver. It is not wise to put the burner near to the gas tank, so provision was made to put everything bar the burner in the box at the rear. The servo for the steering is underneath and the servo that drives the tumbler gears is mounted on the side.

It has forward, reverse and neutral operated by the servo via a rod that moves the gears. When in neutral the engine continues to run and the buggy stands still. When forward or reverse is selected, the tumbler gears are thrown one way or the other to reverse the drive. The drive from the tumbler gears is from a 1in pulley to a 2in pulley and then from an adjoined $^{3}/_{4}$in pulley down to another 2in pulley. This gives a fair amount of reduction. When a Stirling engine is loaded, it lowers the speed of the engine, but the performance improves because the air that is being heated and cooled is in the hot and cold ends for a longer period of time.

A rear-quarter view of the buggy.

On the front and rear there are sacrificial bumpers that protect the buggy. The steering has been set up with the benefit of the Ackermann angle, which optimizes the steering angle of each wheel. It has wishbones on the front with telescopic springs, but no damping. The rear suspension is of a trailing arm-type design and uses a similar coil over rod design. This means that the buggy has fully independent suspension. To simplify the power train only one rear wheel is driven, which eliminates the requirement for a differential.

The gas tank is at the rear in the box with all the electronics and the receiver. A silver-

View showing the black anodized cooling fins.

soldered copper pipe runs from the gas tank to the burner. It has an ordinary gas burner that uses ceramic material inside. The burner is completely encased in an aluminium chimney. The chimney is lined with pieces of plumber's mat, which is a kind of fibre-glass used by plumbers to act as a heat shield when they are soldering pipes. The bottom of the chimney is open, so fresh air enters from underneath. The original side air holes were removed as these caused the flame to blow out when the buggy was first being trialled. The gas tank holds enough fuel for about one hour of operation.

> I once had a problem where a purpose-made plastic fuel pipe blew off under very hot environmental conditions, so now I only use or recommend silver-soldered copper pipe for these joints. This buggy is often run for public demonstration so safety must be paramount.

It has a switch to turn off the batteries. The whip aerial appears to be 1ft long, although the actual length is about 1m but coiled inside the box.

A front view of the buggy.

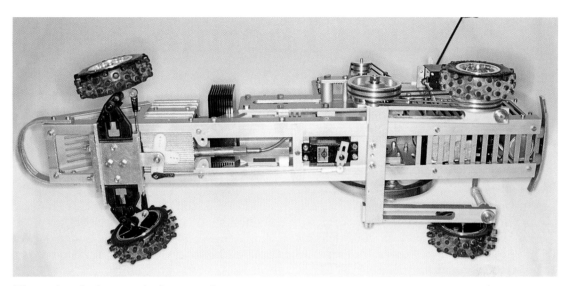

The underside showing the layout and construction.

The square cooling fins on the cold end are very fine and were anodized by my son Graham. By making the fins black, I understand that the efficiency of the cooler is increased by as much as 24 per cent. The buggy gets quite hot in places so when picking it up one has to be careful.

This buggy goes at the same speed as a model traction engine. It took part in the Amberley Chalk Pits Museum traction engine rally, where it trundled along behind eighty traction engines and performed for an hour to the delight of the public.

The host at the Chalk Pits rally said, 'You've seen eighty old engines, now you are going to see a thing of the future: a Stirling-powered Moon Buggy.' It has won 'The best prime mover' (the Salvington Cup) prize awarded by the Worthing and District Society of Model Engineers.

I am fond of this type of engine and intend in the future to make multi-cylinder versions using small cylinders. Any future design improvements will be put on to a new buggy.

14 Lag (Thermo-Acoustic Engines)

The thermo-acoustic engine is also known as a laminar flow engine.

At a show in Bath in April 2002 I learnt about Ted Warbrooke in New Zealand who had managed to get a thermo-acoustic engine running. Up until then I had regarded thermo-acoustic engines as needing to have tubes 30m long, so I had never considered them from a modelling point of view. Within two weeks I had made my own version of this engine, which was probably the first in the UK at that time.

The simplicity of this engine is that it has only one piston, the power piston. This is coupled to a flywheel via a crank. The glass tube with regenerative material in it acts like a displacer. The air inside starts to pulsate when heat is applied and the engine starts to oscillate, eventually self-starting. It runs equally well in either direction.

The frame could be constructed from wood, aluminium, or any similar material. It is a very simple engine; all that is needed is something to hold the power cylinder and the main bearings with the crank disc on it.

THERMAL LAG MK I

I built my Mk I version with a very heavy aluminium base with the idea that the base would be the heat sink and become the cold end, although later I found that this was unnecessary. There is a 6in Pyrex tube, 1in diameter, with a $^1/_4$in tube running along the middle of it. The $^1/_4$in internal tube has stainless-steel regenerative material packed around the outside to fill the whole of the space between the inner and outer tubes. The tube goes into the power cylinder, so you could say

Thermal Lag Mk I (latest layout), flywheel side.

View from the crank disc side.

that the tube was 1 foot long in as much as it is open at the hot end and it comes back on itself, similar to a 'U' tube. It has a cast-iron cylinder and a graphite piston. It has run for hundreds of hours and still runs well. It runs equally well in either direction, at a constant speed of 1,000rpm and apart from occasionally cleaning the piston and cylinder there is no maintenance required.

Other lag engines I have seen have been made without graphite pistons; these work, but not as well. These pistons would require oil, which adds a drag that this type of engine cannot cope with. For this reason, this type of engine really requires a graphite piston and cast-iron cylinder. Occasionally, the dust begins to bind a bit and it needs to be wiped clean.

Ball races are used at either end of the con-rod, connected with a silver-steel shaft. An adjustable crank throw enables the compression ratio to be varied. The main bearing housing can be moved to accommodate the adjustable stroke. The drive mechanism is an ordinary con-rod going straight on to a simple crankpin on a disc.

This engine requires very little heat. The hot end is only hot for a very short distance and the cold end remains cold. Unlike an ordinary Stirling engine, the column of air that moves the piston expands and contracts but does not mix.

After making and experimenting with Mk II and Mk III, the Mk I was changed to a different layout inspired by an old design. The thin tube running up the centre of the Pyrex tube was eliminated. A stainless-steel mesh regenerator was pushed into the end of the Pyrex tube leaving an open area next to the power cylinder. Instead of heating the end or the middle of the tube, the burner was placed closer to the power cylinder and the performance increased markedly.

THERMAL LAG MK II

Thermal Lag Mk II.

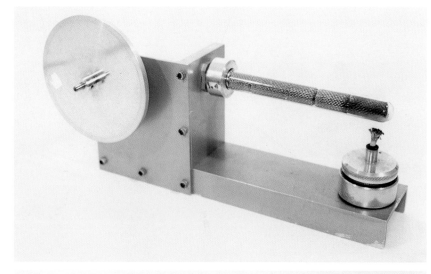

This engine has stainless-steel gauze for the entire length of the test tube.

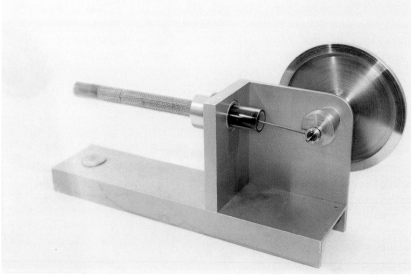

The glass tube and graphite piston can be seen in this view, connecting to the crank disc.

In order to make this engine I contacted Bob Atkins because I needed a piston and cylinder that had a very good fit, and Bob is an expert in this field. Within a few days a glass cylinder and graphite piston arrived from Australia. The Pyrex tube is 6in long and ³/4in diameter with a power cylinder 1¹/2in long and ⁵/8in ID. The piston is 1in long. This engine has been running for hundreds of hours now and the only maintenance needed is to take it apart occasionally and clean the dust out of the cylinder. The engine runs at approximately 1,000rpm.

THERMAL LAG MK III

The Mk III Thermal Lag engine was built to see whether more power could be achieved by increasing the dimensions. Experiments were carried out with a whole series of different diameter and length tubes, up to 1ft long and 1in in diameter, but I found that a ⁵/8in diameter tube, 5in long is the optimum size.

I experimented with stainless-steel hot ends 1ft long and 2in diameter, turning the end white hot, but this would not run either. By reducing the

Thermal Lag Mk III.

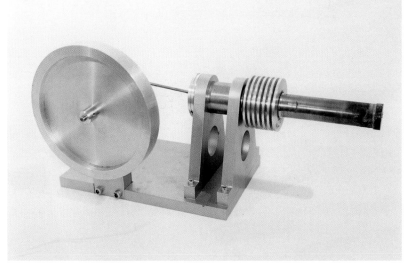

This lag engine is mounted on pieces of a scrap Lancaster. It currently has a metal cylinder and metal cooling fins.

diameter and the length, eventually it began to work and when it got back down to 1in diameter and 6in in length it went really well again. For this reason it will probably remain as an executive toy, 'until somebody proves me wrong'.

Being my first engine of this type I didn't expect a great deal of power in the same way as the first steam engines that were made. Mark III had most of the experiments carried out on it. It had a whole range of different cylinders, displacers, trombone-type tubes and triple tubes side by side all feeding into one piston,

but none of these ideas was successful and after spending a year experimenting with it I decided to concentrate on other things.

THERMAL LAG MK IV (BUGGY)

I trialled this type of engine in a buggy. All kinds of different pistons and cylinders were experimented with, but the power to weight ratio was never sufficient. It would drive the wheels suspended in mid-air, but when it was put on the table the power was not sufficient to

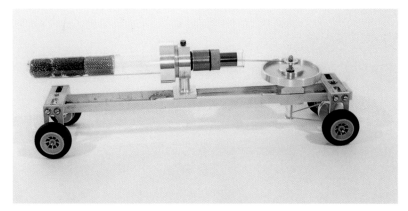

Thermal Lag Mk IV (Buggy).

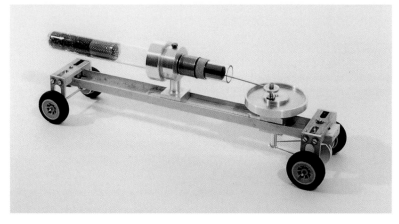

This view shows the horizontal flywheel and the graphite piston in the glass cylinder.

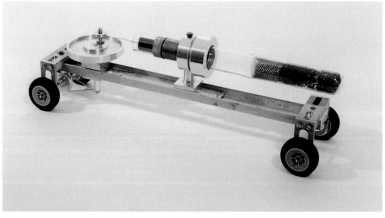

This displacer end view shows the wire gauze and kettle defurrer.

turn the wheels. I returned to concentrating on Stirling engines and on a Stirling buggy instead. The thermal lag engine is on the shelf for the time being. One way to circumvent the weight problem would be to use it in a boat and at some point I intend to make one.

15 Other Interesting Stirling Engines

PETER STANWELL BEAM ENGINE

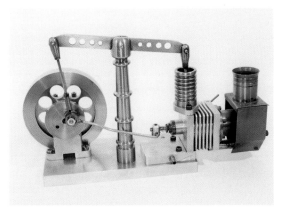

Peter Stanwell Beam Engine.

HOG ENGINE

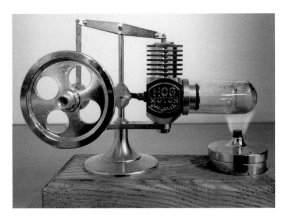

Hog Engine stationary. (Reproduced with the kind permission of Graham Darlington)

This is one of the many engines designed and built by the late Peter Stanwell, who died at the age of 93 in 2003. He was a prolific engine builder and amongst his designs were diesel, petrol, rotary and motorbike engines. I now have the pleasure of owning and caring for thirty of his engines.

This is a self-contained gas-fired beam engine. The power cylinder is mounted at 90 degrees immediately above the displacer cylinder. The engine is made of steel and is relatively heavy, quite unlike the way I make my engines but beautifully machined.

This is a gold-plated engine which I purchased new; I have even seen them in antique shops. They are also available silver-plated. It is an attractive engine that performs brilliantly and this one has run for hundreds of hours at shows.

It has a glass power piston and cylinder and a glass displacer piston and cylinder. Even the guide rod runs in glass. It runs without any lubrication. The power cylinder is encased in a square block with eight cooling fins.

It runs at a high speed and is so successful that it has been used almost constantly at shows, causing the tiny pin bearings to wear.

STIRLING 90 HOT AIR ENGINE

Stirling 90 Hot Air Engine.

GEOFF FORD ENGINE (MODIFIED)

Geoff Ford Engine (modified).

This is a small kit engine with a brass crank-case, power cylinder and cooling fins, which was sold fully machined and ready for self-assembly. I believe that mine was the first produced and it was given to me to assemble and make a report on.

The power piston is 90 degrees to the displacer and it is spirit-fired. Being 90 degrees (the phase angle of most Stirling engines), only one crankpin is needed. The stainless-steel displacer cylinder is held on to cooling fins with four Allen screws. The power cylinder is also held by the same method. The displacer is driven by a small scotch crank and two very thin con-rods made of spring steel go to the power piston. There is a perspex cover on the crankcase so that the con-rod can be seen in operation and I have made a small air hole in it to avoid back pressure on the power piston. A plastic pipe takes air from the displacer cylinder up to the power cylinder and it has a small brass flywheel. It is not ideal for long running as the burner does not hold much fuel.

It has a wooden base and if the engine was raised slightly it would achieve a better performance as it could be fitted with a larger burner, which would also enable it to run for longer.

The designer and maker of this engine was Geoff Ford, a friend who ran a small company that produced limited editions of precision-made engines. He is a very good engineer and craftsman.

The original engine had no cooling fins and only had a small methylated spirit burner. When running it on gas at shows it soon overheated, so extra brass cooling fins were added. I started with the tapered fins, but found that they were not sufficient so added more and this accounts for the irregular shape.

TIN CAN STIRLING ENGINE POWERED BY A NIGHTLIGHT

These engines were photographed at the European Stirling Engine Forum at Osnabrück in 1998. They demonstrate how a simple Stirling engine can be, consisting of a tin can, a nightlight, paper bellows and plywood. The power generated was sufficient to run a radio.

LAWRENCE'S ENGINE

This engine was made by my grandson, Laurance, and is a modified version of an

Tin Can Stirling Engine Powered by a Nightlight.

Laurance's Engine. (Reproduced with the kind permission of Graham Darlington)

engine designed by an Irish friend, Eric Hughes, who called it 'The Advanced All Electric Soup Can Hot Air Engine', because the displacer cylinder was made by soldering two soup cans together and it ran from a light bulb. Eric Hughes' engine has a square flywheel and is made as roughly as possible to show what can be made with whatever materials are available. He gave me a sketch of it and this is what Laurance used to make his engine.

Laurance made this engine at the age of seven and did everything himself except soldering the brass power cylinder on to the displacer and making the loops in the piano wire con-rods. The main frame is plywood and the bearings are wood screws. The displacer cylinder is an aerosol paint tin with a top that he machined himself then glued on with Araldite. Laurance's version runs on a

methylated spirit burner and although it is very crude it runs exceptionally well.

Laurance put this engine into the limelight when he had his photo published in *Model Engineer* (Vol. 181, No. 4,081, 3 December 1998) as the youngest model engineer.

DANIEL LYONNET'S ENGINE

This is a Stirling engine that can be made without a lathe and only using hand tools. Daniel Lyonnet wanted to demonstrate just how simply and easily a Stirling engine can be built, so he made the frame from MDF and the displacer cylinder from a tin can, and instead of a power piston he used a diaphragm. The original diaphragm was made out of material cut from rubber gloves. The thinner the rubber, the better the performance.

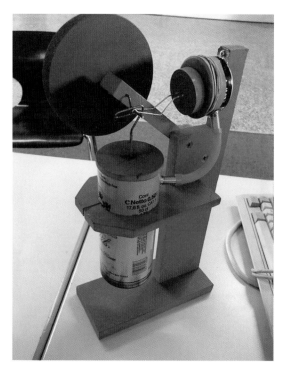

Daniel Lyonnet's Engine.

Viebach Low Δ T Stirling Engine.

Philips Stirling generator frame, flywheel side.

Philips Stirling generator, air intake side.

VIEBACH LOW Δ T STIRLING ENGINE

This is a variation on a Low Δ T, in which the usual flat circular base has been replaced by a soft drinks can that is hand-held. To demonstrate this, a light bulb was inserted into a hollowed-out model hand to supply the heat to the displacer cylinder.

PHILIPS STIRLING ENGINES

Philips Stirling Generator

This is a Philips portable 600W generator, manufactured as one of the few Philips Stirling engine production runs. The frame is pressurized to 25bar. When starting this type of engine the pressure stored in the tubing helps in the start-up sequence. Once the engine has started it repressurizes the frame. It is started with a rope coiled around the centre of the flywheel. It has a regenerator and a forced air burner.

Philips engine.

Philips/DAF bus engine.

Philips submarine engine.

Philips Engine

This is one of Philips' experimental Stirling engines that is vertical and air-cooled. It has a large displacer and a large fan. It is similar to the unit used in the 600W Philips generator.

Philips/DAF Bus Engine

This shows the Philips/DAF engine that was fitted to some experimental buses in the early 1970s. These engines operated at around 220bar. Their power output was 200hp.

Philips Submarine Engine

This is an incomplete Philips submarine engine. It was a forerunner to the Kockums submarine engines.

Philips Cryocooler

The is one of many cryogenic coolers designed and made by Philips. It is for making liquid nitrogen.

Philips Cryocooler.

16 Workshop Practice and Techniques

In my workshop. (Reproduced with the kind permission of Mike Chrisp)

This section covers all the components of a model hot air engine and explores workshop practices as they are relevant to making hot air engines. The intention is not to be a chapter on general workshop practice, even though by definition there will be an element of this, but it is written with an eye to demonstrating the methods I have used and found successful. The aim of this chapter is to fuel your enthusiasm and encourage you to build hot air engines. If you are an experienced model engineer then the hints and tips in this chapter should serve to assist in the quality and performance of these engines. If you are new to these engines then this chapter gives an invaluable insight into the building of Stirling engines.

Always try to use the simplest and easiest method and try to make every component as accurately as possible. Spending a little more time on each component pays off in the end. In effect, spend as much time as is required to make each component and ensure it is made as true and as accurately as possible. The methods recommended in this chapter will help you to achieve this.

The following sections contain descriptions of what I have found to be the most successful ways in which to set about building an engine.

When making a new hot air engine it is good practice to start with the power piston and cylinder. These pistons and cylinders are the heart of a hot air engine. Without a very good fitting piston and cylinder the engine will never realize the level of performance you might hope it would.

CYLINDERS AND PISTON MATERIAL SELECTION

Glass Cylinders
A Low Δ T piston and cylinder is best made using glass for the cylinder and graphite for the power piston. Glass cannot be machined by the amateur model engineer, which means finding a cylinder that is already made and accurate. One of the few sources open to us is to use the glass syringes commonly used by vets. Some of the syringes are quite large in diameter and too

With a suitable glass syringe cut to the required length the two ends need to be smoothed off. To remove the sharp edges and get the ends flat, rub both ends of the glass with wet and dry paper on a flat metal plate. Make sure it is wet and continue until you get a matt surface. If there is a requirement for a smoother finish, a little washing-up liquid may be used and the ends of the cylinder reground. This helps to avoid the small splinters found where the glass is cut.

Graphite Pistons

This Low Δ T Ringbom shows the use of a graphite piston in a cylinder made from a hypodermic syringe.

long for our requirements, so should be cut into sections. The best tool for this is a diamond saw or an abrasive disc.

A graphite piston and glass cylinder.

Take precautions to ensure that you do not breathe the black dust that comes off while machining graphite.

When making a graphite piston to fit a glass cylinder it should not be honed or lapped. Turn the piston at a high speed in the lathe using the newest and sharpest-tipped tool possible; this will give a high finish.

A diamond circular saw used on the lathe set up to cut a Pyrex tube.

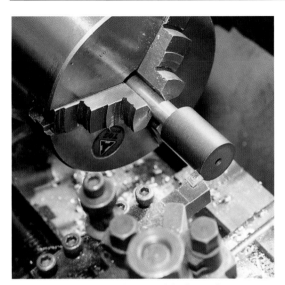

Using a sharp carbide-tipped lathe tool to get a good finish on a graphite piston.

Turn the graphite piston until it just enters the cylinder, using the cylinder as a gauge. Make half a dozen or more passes along the piston at the same lathe cross-feed setting; this will only remove a tiny amount of material, but will give the best finish possible. When the piston just enters the cylinder this is the time to stop, and part it off.

The cylinder and the piston will be a gentle push fit and the piston can be worked in and out until the fit is achieved, but remains gas tight. That is about the best method I have found to make a graphite piston fit a glass cylinder.

> The graphite dust falling into the lathe is not a problem because it helps to lubricate the bed, graphite being a good lubricant. You will find that the bed becomes a lot smoother once you have been cutting graphite. In an old ML7 lathe that I had for forty years, I even packed graphite dust into the top slide lead screw.

Cast-Iron Cylinder with Cast-Iron Piston

The ideal combination for a model is a cast-iron cylinder, with a cast-iron piston; this is partly because of the graphite or carbon in the cast iron. With a good quality cast iron, there is a fair amount of graphite in it. This combination can run together without much, if any, additional lubrication. When adding lubricant, put a trace of oil on the piston and it will run well for many hours without any other lubrication.

Aluminium Alloy Cylinder with Aluminium Alloy Piston

A light aluminium alloy cylinder bored from a good quality aluminium alloy with an aluminium alloy piston is another good combination. It should be accurately turned, honed, lapped and polished. I recommend putting a light anodizing on the piston. This achieves a soft piston, highly polished, that has a hard-wearing surface of aluminium oxide.

Experience has shown that if the piston has actually been run in the cylinder prior to anodizing a very good fit can be obtained. This is because the piston increases in size by a few microns after anodizing.

Lead Bronze Cylinder with Aluminium Alloy Piston

A selection of cylinders made from: (clockwise from top) brass with cooling fins; lead bronze; graphite; stainless steel; glass and aluminium alloy.

A selection of pistons made from graphite, aluminium alloy and leather.

Lead bronze with aluminium alloy is hard-wearing and tough. The cylinder is heavy and dense, which give it good vibration and noise absorbing properties. When fitted with an anodized aluminium piston it will form the basis of a smooth, quiet, long-lasting and easy revving engine. The lead bronze has a good appearance when polished.

Brass and Cast Iron

This good combination will also last for years, has been tried and tested and has good lubricating properties.

Leather

A small leather cup washer that forms a lip seal makes a gas-tight piston that is usable in a light alloy or a brass cylinder. It operates in a similar manner to a cycle pump. This makes a very good gas-tight piston, and can improve the performance of hot air engines where the piston and the cylinder may not have been a very good fit in the first place. The assembly can also prove to be considerably lighter than an equivalent piston and con-rod. Leather pistons work well in any metal or glass cylinder, but will require occasional lubrication to prevent them drying out.

A lightened piston made from leather.

> Remember, a soft material will always wear a harder material. The softer material can pick up hardened particles, which causes this wear.

Rulon

Rulon may also be used as a lip seal in a similar way to leather. There are many different grades of Rulon that are appropriate for different cylinder materials. *See* Appendix II for more information, and visit the website www.stirlingengineer.com for further details and data sheets.

CYLINDERS

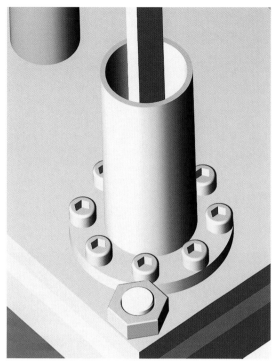

A typical power piston flanged cylinder.

A plain cylinder with screw-cut male thread.

Decide before you start if the power cylinder will need a flange, as this will be used to bolt or screw the cylinder on to the main crankcase.

If a flange is used, then it should be completed before the cylinder is bored. Using a tube in the relevant material close to the intended size, machine or silver-solder a flange on to the cylinder. This should be completed prior to any other machining operations because of the likelihood of distortion when silver-soldering.

If the cylinder is to be threaded into the crankcase, rather than flanged, the cylinder should be threaded first and used as a gauge to thread the crankcase. Use the male thread to gauge the cut of the female thread. It should then be screwed in against an O-ring. If no O-ring is used I would recommend sealing it with Hermatite on the thread.

Boring the Cylinder

Do not be overly concerned by the specific tolerances of the bore, as long as there is sufficient material to bore to achieve the size required. It is more important to get a perfectly true parallel bore with a good finish rather than specifying an actual diameter or tolerance.

A sharp tool will be required (in general all your tools should be kept sharp) and use a fine feed on the lathe. When you are satisfied with the bore, pass the boring tool through many times without altering the cut, to make sure that you have taken all the spring out of the tool, because when you bore a cylinder it will always be a little bell-mouthed and will be ever so slightly larger at the end where the boring bar enters than at the chuck end.

Honing the Cylinder

I would recommend using a honing tool to hone the cylinder. Details appear at the back of this book of companies which sell these types of tool. They are very similar to the honing tools that are commonly used in the manufacture and repair of internal combustion engines in the motor industry, but remember that a hone follows any discrepancies in the bore. If the original bore was slightly triangulated, the lobes on the hone tend to follow the shape of the

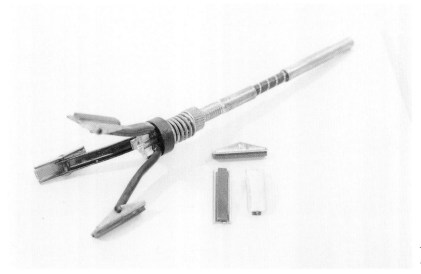

A honing tool and spare stones.

bore. It does not therefore necessarily make a true hole. The idea is not to use the hone to make or improve a poor hole; rather, we should concentrate on boring the hole more accurately in the first place. A hone should be used only to remove tool marks. If the boring has been done well, with a brand new tool and a fine feed, there should be very few tool marks to remove.

The cylinder should be removed from the chuck and replaced by the hone. A little oil is

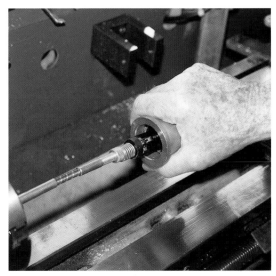

Honing.

Removing the tool marks.

recommended when using the hone. Hold the cylinder, close the three stones up inside the cylinder and with a low speed, physically move the cylinder up and down the hone to remove any tool marks. You get a criss-crossed pattern, which, if this were for a steam engine, would be quite beneficial to leave in the cylinder to hold the oil. However, in a hot air engine a further operation is required: lapping.

A wooden lapping tool.

The split on a lap should be about three-quarters of the length of the lap.

If you have no access to a hone a wooden lap can be made, and by using a very fine grinding paste you will have an effective honing tool.

Lapping the Cylinder

Turn a wooden lap to fit the cylinder, drill a small pilot hole to a depth of about 1in and then saw down the middle along three-quarters of its length and insert a wood screw into the end to open up the split. This may not seem to be a very accurate instrument because you are

opening up one end and the other end cannot possibly be the same diameter. However, you will find that this does polish the cylinder and after a little application of working the cylinder up and down on the lap, it becomes parallel as it wears and you can polish the cylinder using this wooden lap with ordinary Brasso-type metal polish.

There is a material available called Time Saver Lapping Compound, which is a very fine

Offering the cylinder up to the wooden lapping tool.

grade that is ideal for polishing hot air engine cylinders; it breaks down differently to the traditional carborundum grinding compound. It gives quite a high polish to a cylinder. Some of the carborundum compound and powders are inclined to retain their cutting ability and don't in my experience polish anywhere near as well. You should aim for a highly polished, accurate finish to the cylinder.

> Laps that have been used with lapping compounds should be stored in clearly marked containers away from the ones that have been used with polish, as cross-contamination can occur. Laps may be kept and recycled for smaller cylinders. Make sure that the lathe bed is covered when turning them down to size to prevent abrasive compounds finding their way on to the lathe bed.

Summary

- The boring should give a true cylinder.
- The honing should be used to remove the tool marks.

- The lapping should be used to achieve the ideal surface finish.

The cylinder should then be washed thoroughly with washing-up liquid and rinsed to remove any traces of grinding material. Dry it as soon as possible with kitchen towels and then set it aside to wait for the piston.

TOLERANCES AND FITS

Make the piston to fit the cylinder, rather than worrying about specific fits. Concentrate more on an optimal gas-tight fit with minimal friction.

In my experience of working on Rolls-Royce Merlin engines in Spitfires and Lancasters amongst others, I believe the reason why Rolls-Royce earned its early reputation for quality engines is because pistons were fitted to each individual cylinder. When making an individual hot air engine you have the opportunity to do just that.

RAF ground crew working on Merlin engines.

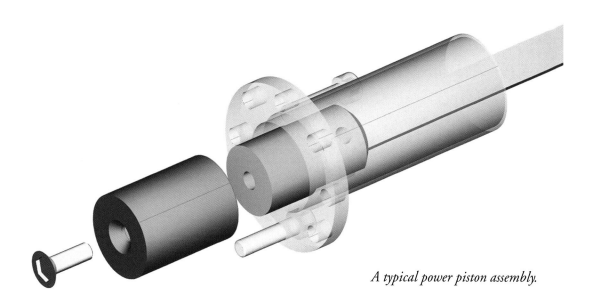

A typical power piston assembly.

POWER PISTONS

The power piston should be gas-tight, with as little friction as possible. A clevis is attached to the piston via means of a countersunk screw through the crown of the piston. For hot air engines of relatively low power this method is preferable as it eliminates having to cross-drill the piston and reduces the possibility of the gudgeon pin rubbing against the side of the cylinder. It improves the longevity of the cylinder, and helps to reduce the leakage that could occur via the hole for the gudgeon pin.

It is preferable to have the outside of the piston wall unbroken all the way along. For some applications, a shallow oil groove can be machined, although this would normally act as a gas 'stepping stone', causing leakage similar to a hole for a cross-drilled gudgeon pin. It is typically placed closer to the crown.

Boring and Drilling the Piston

When using a metal piston it should be made as light as possible by thinning the walls internally. Drill right down the centre of the

An example of a piston with oil control slots and a gudgeon pin hole.

piston towards the crown with a clearance hole of relative size to the thread that will hold the clevis. Next, the internal part of the piston is counterbored. The thickness of the crown should be sufficient to accommodate the head of the countersunk screw that will hold the clevis.

The wall thickness should be about $1/16$in, but when boring the internal part of the piston leave $3/32$in of material so that when the external side of the piston is machined the wall of the piston ends up at approximately $1/16$in thick.

Turning the Piston

Turn the piston until it is nearly down to the size of the cylinder. Reduce the piston to within a thou of the bore of the cylinder. Then use the external hone to take off the final half thou or so. Achieve the final polished finish with a lap.

An external hone in operation.

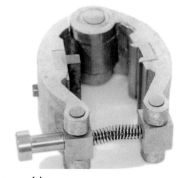

An external hone.

A piston being carefully offered up to a cylinder.

Once the piston enters the cylinder it may have a tight spot where it can only be pushed through so far. Very carefully, with a little metal polish, push the piston in and out of the cylinder. High spots will be identified and the piston is effectively using the cylinder as a kind of lap. Work it in and out of the cylinder so that the piston slides nicely through the cylinder and achieves a high polish on both piston and cylinder. However, if this operation is overdone, the piston for this cylinder can be ruined. Make sure that cylinder and piston are spotlessly clean, dry and oil free.

Checking that the piston and cylinder fit.

Testing a piston.

A distorted piston.

The trial to determine whether the fit of the piston in the cylinder is satisfactory is to put the piston in the cylinder and block the air with the palm of your hand or thumb. The piston should not move. When you remove your hand from the cylinder, allowing the air to get into the cylinder, the piston should be able to drop right the way through.

Make sure though that if the piston should drop out that something soft is made available for it to land on as they are easily damaged.

If making an aluminium piston use the same procedure. Afterwards run it for an hour or so in the engine, then take it out and anodize it. After it has been anodized you may find that it does not go back into the cylinder as it has increased in size by a few microns over the diameter. It may form a peak or two during the anodizing, so a gentle polish with a wooden lap will remove this. Anodizing makes a really hard surface and it will ensure a long-lasting piston that is also nice and light.

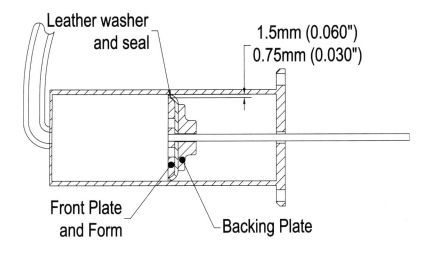

Leather washer and seal

1.5mm (0.060")
0.75mm (0.030")

Front Plate and Form

Backing Plate

The die for the leather washer should be reduced to allow the appropriate clearance.

117

Making Leather Pistons

Leather pistons are generally very good pistons and provide a nice gas-tight fit. Use a chrome leather, as found in cheap gardening gloves, as these have a sort of chamois leather on the back of them, which being very soft leather, is ideal for use as a piston.

When making the leather piston, the aim is to have a small lip in contact with the cylinder. Initially, however, there should be plenty of excess material, which will be trimmed off at the final stage.

The leather piston assembly, comprising from left to right: the die washer; leather piston; and the top hat section clamp.

An untrimmed leather piston in a jig.

Measure the diameter of the bore of the cylinder, then, using a micrometer, measure the thickness of the leather, which is usually between 0.030 to 0.050in. Next, make the metal washer that the leather will be formed over. This washer should have a minimum radius of 1× leather thickness cut on the side nearest the leather.

If the thickness of the leather is 0.030in make sure that the washer that forms the inside of the piston is at least 0.060in smaller than the bore of the cylinder. With a different leather

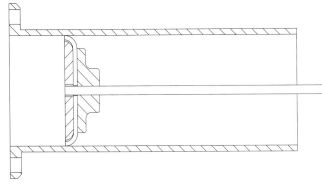

Sectional drawing showing shaft, washers and leather piston as it sits in the cylinder. Note the radius on the die washer that the leather is curved around.

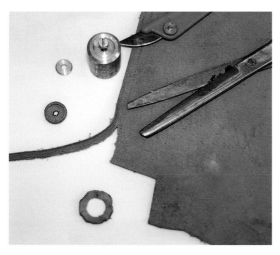

The materials and tools used to make the leather piston.

The piston and jig assembly after trimming.

thickness it follows that the washer should be sized using the same ratio. The inner washer has a fillet radius machined on to one side at the edge. Both the inner and outer washers are threaded on to the shaft. The two holes in the outer washer are to tighten the washer when assembling the parts. It is advisable to ensure that the outer washer, the shaft and the leather washer remain flush at the end of the cylinder when assembled. This prevents any protruding parts hitting the end of the cylinder and minimizes the dead space of the engine.

The leather should be soaked in water first, then clamped between the inner washer and the outer washer. Force the washer into the cylinder and leave it to dry, preferably overnight.

A leather piston being trimmed with a craft knife.

The new leather piston and the jig disassembled. The aluminium washer in this instance will form part of the final piston assembly.

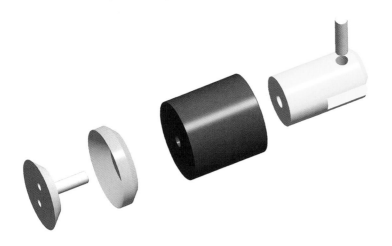

*Exploded view of a lip seal
and piston assembly.*

Remove the newly formed piston; it will have taken on the shape of a cup washer (as used in cycle pumps). Place the newly formed piston back into the cylinder. With the face of the internal washer flush with the end of the cylinder, cut off the excess leather with a sharp craft knife, using the end of the cylinder as a guide. This may also be done using a specially made jig. The excess material should be trimmed so that you do not end up with a very wide piston. Aim to make a small lip seal

It should then be soaked in oil and this will make a wonderful piston that will run for years, provided it is kept moist with oil.

Rulon lip seal, jig and piston.

Rulon Pistons and Seals

It is possible to use Rulon sheet to make a lip-type seal. The Rulon should be formed, heated and cooled in a jig. This assembly should be heated for approximately thirty minutes at 200°C (400°F) and allowed to cool for a further thirty minutes.

Rulon, however, is not ideal for a piston in a hot environment due to its large coefficient of thermal expansion. The problem if using it with a piston is that if it is made with sufficient clearance to allow it to run when hot then the engine will be almost impossible to start due to the poor seal. However, if it is made to allow for

this then the engine will not run very well, if at all, due to expansion causing excessive friction.

Rulon should therefore only be used in hot air engines in very small quantities and thicknesses. As a lip seal, it expands, giving a very fine feather-edge that engages the cylinder and has very little friction. These lip seals make a good seal and give an excellent performance. Rulon's lubrication is very good and does not require oil. Its exceptional low friction properties make it absolutely ideal for using as piston rings or as a sliding bearing surface. (*See* Appendix II for the properties of this and other materials.) Well thought-out uses of Rulon can transform the performance of a Stirling engine and development of materials such as this can only serve to secure, enhance and broaden further the applications for which Stirling engines may be used.

DISPLACER CYLINDERS AND PISTONS

Matching a displacer piston and cylinder.

Cutaway view of displacer piston and plug.

Displacer cylinders, sometimes called hot caps, and their pistons are the parts of the Stirling engine that initiate the process which allows the energy in the form of a heat differential to be converted into shaft power. The loose-fitting piston that fits inside this cylinder is the displacer piston. These items demonstrate better than many other processes how the laws of thermodynamics and gas can interact to give us useful work from heat. An overview of these laws can be found in Chapter 2. It would be useful to re-read these laws and consider how they are applied in the way the displacer cylinder and piston function in a real environment.

The most appropriate material for the displacer cylinder is stainless steel. A usable cylinder can be made of mild steel, but this is not desirable, partly due to corrosion. Stainless steel exhibits very good properties at elevated temperatures. The heat conduction of stainless steel through the length of the displacer cylinder is not good, therefore it makes an ideal material for the displacer piston and the displacer cylinder. Titanium has a similar level of heat conduction and some grades have half that of stainless steel.

In a Stirling engine, it is necessary to get one end hot and one end cold and the greater the differential between these two extremes, the better the efficiency and thus the performance of the engine.

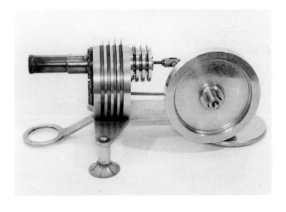

Note the blackened hot end and the fins on the cold end.

The displacer cylinder and piston were once available commercially as matched pairs but now these are not readily available, which is a great pity as you could select the size you wanted and you had a displacer piston that was a few thou smaller than the displacer cylinder it slides in.

A common misunderstanding of the way the displacer piston works is that all pistons should be gas-tight. In order for the displacer piston to perform its function it must not be gas-tight in a standard hot air engine. You have to have a gap all round the displacer piston to allow the air, or other working medium inside, to move between the hot and the cold ends of the cylinder chamber.

The wall of the displacer cylinder should ideally be about 0.010in thick, as this means there is less cross-sectional area of metal for the heat to conduct along its length. The thinner the walls of this cylinder the easier it is for the heat to go through and heat the air inside the hot cap.

The machining of the displacer and cylinder takes a long time. Free-cutting stainless steels are not to be shied at. Machining grades of

stainless steel such as 304 are readily turned and bored, enabling you to get down to a 0.010in wall thickness relatively easily. It is one of those jobs that feels like there is nothing left by the time it is finished, except a big pile of swarf.

There are various pepper or salt pots on the market that are made of stainless steel. Some have ends pressed into a tube. These can be turned down, which saves a good deal of work, and although the ends have been pressed in, I have used quite a few of them and found that they do not leak.

There are also pressings available, such as sugar bowls and milk jugs. If you look round the kitchen departments of some stores you can find stainless-steel containers that slide inside other stainless-steel containers. If you are not too bothered about the eventual size of the engine, you can make the engine according to what is available. Another way is to buy thin-walled stainless-steel tubing so that there is not so much to turn. Some shock absorbers contain suitably sized tubing, but these tend to be made from steel, which, though usable, is not as good as stainless steel.

A range of condiment containers that are of a suitable size and material for displacer pistons and cylinders.

Sugar bowls suitable for use as water jackets.

Machining a Displacer Piston and Cylinder

This shows the gap between the displacer piston and cylinder.

Turn the outside diameter of the piston to suit your requirements. The hot end of the piston needs an end brazed on to seal it. The opposite end requires an aluminium plug. Before turning down the tube to make the displacer piston, an allowance of 0.050in to 0.060in for a 20mm to 60mm bore across the diameter of the tube is required between the outside diameter of the displacer piston and its cylinder. The piston's wall thickness should be 0.010in to 0.030in.

To turn the tube for the displacer cylinder, cut a piece with enough extra to allow it to be held in the chuck and bore the inside to the diameter required. Then thin down the outside of the cylinder to leave approximately 0.020in material thickness. Bore the inside of the cylinder to achieve the correct wall thickness, then part it off to the length required, allowing a few thou for cleaning it up and facing the ends. The thin walls of the tube mean that you cannot successfully hold it in a three-jaw chuck because it will become distorted with the pressure of the jaws. A way to achieve this is to make an aluminium split collet for the facing and cleaning it up.

To make an aluminium split collet, internally bore a tube to whatever the outside diameter of the tube that needs to be machined to and then split it along its length. The length of the aluminium collet should be about half the length of the piston, allowing some extra

A split collet.

Once collets have been used it is worthwhile storing them and remachining them for use at a later date for other cylinders.

A similar operation to this is required to make one end of the displacer piston. Once again it should be brazed. Reduce this displacer piston tube also to about 0.010in wall thickness to reduce the weight. It is very important to keep the weight of the displacer piston as low as possible.

An alternative method to make the cylinder or piston end plate is to braze it to a piece of stainless-steel sheet and trim the sheet later, but I prefer to turn discs and fit them as above.

length on the collet to enable it to go into the chuck.

The split collet is placed into a three-jaw chuck, with the split between two of the jaws, and the displacer cylinder tube is then placed into it. The displacer cylinder tube should run true. A centre punch mark should be made on the outside where number one jaw is so that the collet can be replaced in the chuck in the same position to maintain as much accuracy as possible. Bore it along its length until the wall thickness is around 0.010in. The mouth of the tube should be ever so slightly bell-mouthed in order to allow for brazing the end into it. I turn a thin disc and part it off, tap it into place and flux it ready for brazing.

Above *A braze coiled around the base of a cylinder ready for brazing to a flange.*

Left *A bell-mouthed tube.*

Despite the flame making the cylinder red hot, the flame is invisible in sunlight. It is also difficult to see the subtle variations in colour, which help identify the temperature of the component.

In order to braze, put a ring of brazing rod sprung inside the tube and coat the end of the tube liberally with borax-type brazing flux. I recommend mixing brazing flux with methylated spirit into a stiff paste rather than water, because water turns into steam and it often bubbles and moves away from the surface. Methylated spirit evaporates quicker than water, doesn't bubble into steam and gives a neater job. When using flux mixed with methylated spirit for any brazing job, you should put the flux around the edge of the tube and heat it up, heating the whole tube end. When the melting temperature is reached, the brass will flow into all the places in one go. This technique is applicable for silver-soldering as well as brazing.

Silver-soldering is not really suitable here for the cylinder and should be considered only for use with a methylated spirit burner.

When brazing, and the same holds true for silver-soldering, leave the job until the evening when the light begins to fade. Usually when I wish to braze something it is always brilliant sunshine, but it is worth waiting until the light fades, so that you can see the flame; when the metal begins to go cherry red that is when the braze will begin to run. It is most helpful to see that you have an even temperature. The golden rule is always to make sure that the surfaces to be joined are spotlessly clean. Even if the piece was cleaned the previous day, make sure it is cleaned again just prior to these operations. As soon as the braze runs, shut off the torch to achieve a neat job.

Centring the indexing tool between the chuck and the tail stock.

125

This gear wheel allows a multitude of indexing solutions. The yellow marks are the ones I usually use.

INDEXING HOLES

If a disc was brazed into the end of the displacer tube, use a mandrel to mount it on the lathe and gently remove any surplus brazing material. It also gives an opportunity to true the end up if it has become slightly out of line after heating. The mandrel should have a slight taper of about 1 degree to enable the hot end to just slip over.

For simplicity, a flange should be brazed on to the other end of the displacer cylinder. Ideally the flange should have a sufficient number of small holes for screws to bolt it on to the crankcase with a gasket in-between. One way to ensure that the holes are in the correct position on the PCD is to use an indexing tool.

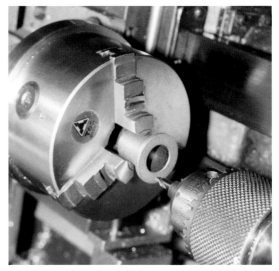

Lining up the centre of the PCD to drill the holes at the registered index point.

I prefer to cut an internal screw thread in the crankcase and an external thread on the cold end of the displacer cylinder. The external thread of the displacer cylinder should be cut while it is on the mandrel. I have successfully screw-cut cylinders that are only 0.020in thick and managed to have sufficient thread on them. Make sure a fine thread is used and use Hermatite on the threads to ensure a gas-tight fit where it is screwed directly into the crankcase. If one doubts their screw-cutting ability a flange should be favoured for simplicity.

The displacer cylinder screwed into the lead-bronze power cylinder that is inside the copper water jacket on Stirling Silver II.

POWER PISTON MANUFACTURE

Always machine the outside of the piston first to make it fit the cylinder. If the piston is not the required fit, there is little point in continuing to work on the inside of the piston. If you do make a mistake many hours of work would have gone to waste. When nearing the final cuts to fit the piston to the cylinder, just pass it half a dozen times to make sure the spring is taken out of the lathe tool.

A lot of honing and lapping of the piston should be avoided. The cylinder must be made as accurately as possible and as the lathe turns perfectly accurately both cylindrically and in parallel, it would be better to leave the piston truly turned, rather than lap it to the point where it becomes barrel shaped.

When lapping or honing, the edges of a short piston always tend to wear before the middle and cause the piston to be slightly barrel shaped. It is therefore preferable to get as good a finish as you can, just entering the cylinder, then with a very light polish it should slide through the cylinder.

Use a small drill through the crown of the piston first, then open it up with larger drills, until a boring bar can be used. Then bore it to the required depth. I would recommend using a digital read-out to get the depth; these read-outs can be obtained quite reasonably these days and make life a lot easier.

After putting the relevant clearance hole in the middle of the crown of the piston, countersink it. Use a proper HSS (High Speed Steel) countersink, which is the same angle as the head of the screw. The countersunk screw that attaches the clevis should be sealed with Hermatite. It seals a bit like a valve and makes a perfectly gas-tight joint. Put a little paper washer underneath the clevis so that when it is tightened it becomes airtight. The inside face of the piston needs to be flat and machined right across. On the final cuts take the wall right down to approximately 0.050in purely to reduce weight. There is no stress on the wall

An external wooden lap.

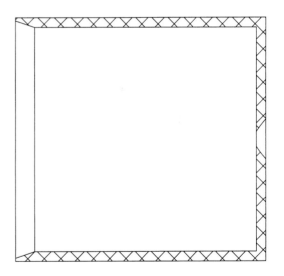

Cross section of a typical power piston.

of the piston so it is quite possible to go down to 0.030in if required.

Cutaway of a piston assembly.

A view of an exploded assembly from the underside of the piston.

A typical piston assembly,

Once the piston has been machined, put a smear of Bluebell brass polish on the piston. I have made a small knurled handle that screws into the piston. This enables me to hold on to it during the polishing operation.

Screw the piston on with a rubber washer in order to make it airtight and sit for half an hour or so working the piston up and down inside the cylinder. You should be able to feel if there are any high spots on the piston while working it up and down. It should be airtight at this stage, so seal the cylinder with the palm of your hand to feel the compression as you polish.

Once satisfied with the fit, the piston and cylinder should be thoroughly washed. Put a smear of oil on the piston and make sure it is going to be a good fit, then wash it all off again before the piston goes for anodizing. After anodizing, the piston will be glass-hard and left in its natural colour; this makes for a very hard-wearing piston. The glassy and slippery surface is very different from plain aluminium. An anodized aluminium piston running in a bronze cylinder should last for years.

The friction goes down considerably once the piston has been anodized. It may be necessary to polish the piston again when it returns from anodizing as it may have increased in size slightly. If this is the case, put a lick of Brasso on with a wooden exterior lap and after a few strokes it will enter and make a lovely fitting piston. Don't be put off if you do not have the facility to anodize, because even without anodizing the piston will work merrily away for many years.

CLEVIS

The clevis changes the angle of the force applied to the piston to allow it to line up and drive through the relative angle of the journal of the crankshaft.

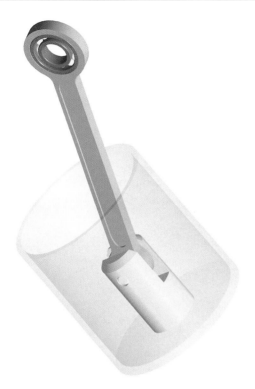

An assembled clevis through a clear piston.

A collection of various clevises.

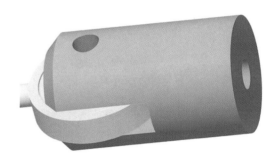

A con-rod and clevis assembly.

Before starting to make a clevis you need to decide what type of bearing the little end is going to have. Normally my little ends on the con-rods are ¹/₈in thick aluminium with a bronze bush in them and a ¹/₈in silver-steel gudgeon pin going through the bush. If a ball race is used in the little end, ensure that the clevis will accommodate the ball race support diameter. The width of the slot in the clevis should accommodate the width of the ball race or con-rod plus a small amount for a suitable clearance. This clearance should be between 0.004in to 0.015in across the slot.

From the centre of the small end, double-check the radius required to make allowances for the bush. I often turn a clevis from ¹/₂in diameter aluminium rod, machining it all in one go without removing it from the chuck if at all possible. This keeps it nice and true.

Use a collet in the lathe to hold the material. There is usually no need to machine the outside diameter. Bore a hole through the centre ready to tap the hole at the bottom end. I use a small 3in three-jaw chuck to hold the ¹/₂in material. When the hole is bored and the end faced, a tapping hole is drilled right through the clevis to the required depth.

Before removing the work piece and the chuck off the lathe, use a small 'V' tool to put an accurate scribe line around the clevis where the gudgeon pinhole will be. This 'V' groove will locate the tip of a small centre drill at a later stage. Remove the chuck from the headstock with the clevis still in it and put the chuck on the indexing attachment. Square it up 90 degrees to

The pre-prepared blank.

Centre-drilling the end of the clevis.

the centre line of the lathe and mill the slot for the con-rod to the depth required by winding the cross slide in. This can be done with a narrow slitting saw. Work from the middle outwards by revolving the spindle of the milling attachment 180 degrees. This method ensures that the slot remains right in the middle of the gudgeon pin. Move it forward and put a centre drill in the chuck in the headstock and centre-drill across the clevis for the gudgeon pin, then drill to just under the desired measurement. Ream it to the correct size and push the gudgeon pin in so that it is a tight fit and does not move. As the clevis is inside the piston, the gudgeon pin could be left

long. It can't interfere with anything because it can't get through the walls of the piston.

To lighten the gudgeon pin I sometimes put a drill down through the middle of it, which decreases the weight and it becomes a silver-steel tube that is already ground on the outside.

The slot for the con-rod and the hole for the gudgeon pin are all machined without removing the clevis from the chuck, which keeps it all perfectly square. All that remains to be done is to part off the clevis and clean it up.

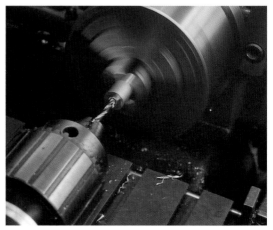

Using a tapping drill to bore the hole in the end of the clevis.

Tapping the hole.

Facing the end of the clevis.

Marking a centre for the pivot pin hole using engineer's blue and odd-leg calipers.

Cutting a chamfer on the clevis.

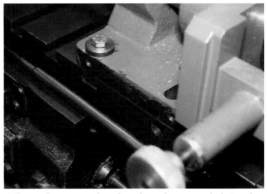

The indexing tool is fitted to the cross slide and squared up as near as possible by eye.

Right *Centring and zeroing an indexing tool with a DTI on an accurate bar held in a collet on the cross slide.*

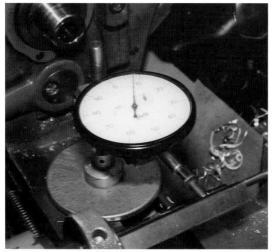

Measuring the inaccuracy of the run-out of the index mounting. This should be zero for its entire length.

Boring the pivot pin hole through while the clevis is held in a collet.

Centre-drilling the pilot hole for the pivot pin.

Above right *Reaming the pivot pin hole.*

Right *Cutting the slot using a slitting saw.*

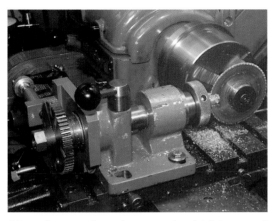

View showing the indexing tool, which is turned through 180 degrees to ensure a perfectly symmetrical slot.

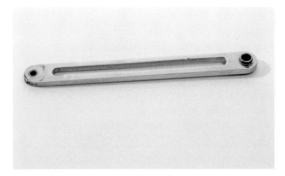

A simple con-rod with plain bushes.

The finished clevis.

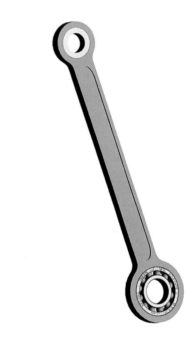

CON-ROD

I will describe how to make a plain con-rod cut out of $^1/_8$in sheet aluminium fitted with plain bronze bushes.

Work out the distance required for the two holes that will house the little end and the big end. A plain bronze bush in the little end is quite satisfactory as these last for many years. If a nice reamed fit is used on the silver steel it should be quiet. Allow a little side clearance on the clevis to allow for any slight misalignment of the con-rod. However, this should not strictly be necessary.

A bearing and bush con-rod assembly with material removed to optimize cyclic stress performance.

I often make my con-rods parallel and just mill a slot up the centre to reduce the weight. Machine the aluminium blank to the thickness of the ball race. Then drill a hole for the little end.

In order to mark out the hole for the con-rod I usually use the milling machine with a digital read-out to get the two holes accurately placed, although this is not essential because a thou or two either way doesn't really matter. It can equally be done on the drilling machine. Mark out where you are going to put the gudgeon pin hole and put your centre punch in there.

For manual marking out use engineer's blue, a digital caliper/vernier and scribe a line along the middle. In aluminium use the calipers to mark the metal. Put a centre-punch mark where the first hole should be. Before centre-punching anything use a sharp scriber to find the centres. Scribe cross lines and pick the centre out with a fine pointer and press that in by hand, then centre-punch.

Using dividers, set them to the required length. Place one point in the first centre-punched hole and scribe an arc across the centre line. Put a pinpoint mark where it crosses. Now centre-punch as before and drill both holes.

To mark out on a mill, also paint engineer's blue on to the aluminium blank.

Scribing the centre line.

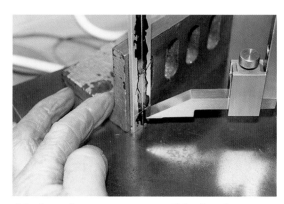
Marking the centre position of the first bearing hole.

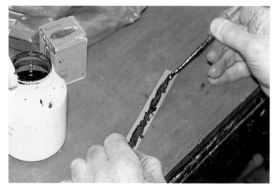
Using engineer's blue.

Hold the aluminium blank against a block that is square to a surface plate and scribe a line using a digital vernier.

Mark the centre position of the first bearing hole.

A wiggler tool can be used to find the position of the edge of one of the jaws of the vice. When it stops wiggling it shoots up to the position shown, thereby determining the precise position of the edge of the vice jaw.

The diameter of the wiggler ball is measured, half of which determines the true centre of the

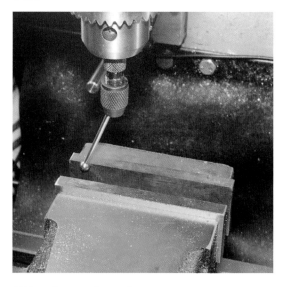
Using the wiggler tool.

milling spindle. The cross feed is then repositioned to half the diameter of the wiggler ball.

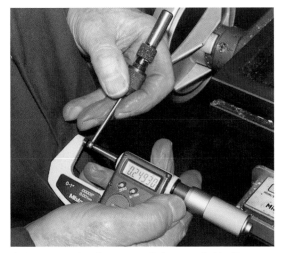

Measuring the diameter of the wiggler ball.

Measure the width of the metal. Half of this measurement determines the amount that the cross feed on the mill should be moved to find the true centre of the metal. Depending on the tools available, it may not be necessary to mark the metal manually, thus increasing the accuracy. The metal should be centre-punched and if you have access to a Jeweller's Loup (cross-hair sightglass), use it to retain the accuracy of the centre-punch mark.

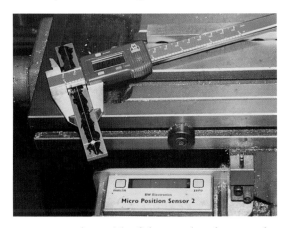

Measuring the width of the metal and setting the cross feed to half of that measurement.

Now centre-drill the hole; if you have followed the previous steps you can be confident that the centre is correct.

Centre-drilling the first hole.

Drill the other bearing hole at the specified distance.

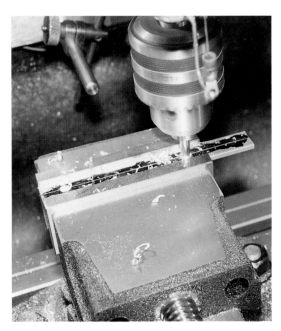

Drilling the second bearing hole.

Without removing the metal from the vice, mill the lightening slot. The accuracy of the slot is now easy to maintain. The metal can then be removed from the vice. The ends should be shaped. All edges should be broken (deburred) and the bearings should be mounted in accordance with the desired method of mounting.

Milling the lightening slot.

Put the con-rod on the faceplate with a little packing behind to protect it. This stops the con-rod being pushed into the hole in the centre of the faceplate, also protecting the faceplate at the same time. Bore a hole for the ball race. I used to bore holes for ball races so that they were a tight press fit, but small ball races can be secured quite satisfactorily with a smear of Loctite to hold the bearing in. Bore the hole for a sliding fit, allowing 0.004in across the diameter to accommodate the Loctite and provide the optimum bonding gap. Let the Loctite cure as per the instructions.

Drill and ream the hole for the bush on the drilling machine. If using a $^1/8$in gudgeon pin you can use down to a $^3/8$in bronze rod to make the bush. Drill and ream the hole to fit the diameter of the gudgeon pin and while still in the lathe part it off to $^1/8$in thick. Use the drilling machine as a press to fit the small end bush in position. With the chuck closed, guide the bush into place, having smeared the edge of the bush with Loctite beforehand. This keeps the bush square and flush with the thickness of the con-rod. Wipe off any Loctite from the con-rod and drilling machine table as soon as it is in place.

The ball race and con-rod on a Low Δ T.

Prior to putting the bushes and the ball race in, mill the centre away to reduce the weight.

If a ball race is required at both ends, put a piece of bar in the lathe and make the two bushes to hold the ball races, one for the little

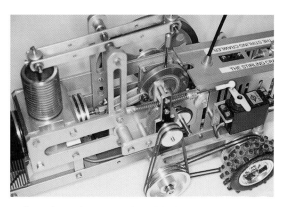

A bicycle spoke con-rod and ball race used in the Stirling-Powered Buggy.

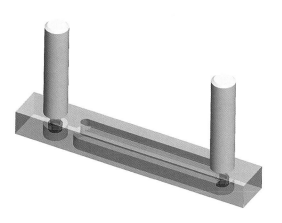

This adjustable jig is required to assist in the accuracy of making multiple con-rods for a multi-cylinder engine.

end and one for the big end. Turn them in the lathe and then centre-drill from the edge. Put a silver-steel or stainless-steel rod in to make the con-rod. Work out the distance between the two con-rod holes and put two silver-steel pins in a jig so that when the three parts are assembled, the stainless-steel rod with a little Loctite can be put into the cross-drilled hole. Put the assembly on the jig and bond the other

This shows how thin a con-rod can be.

end with a little Loctite. Slide it over the silver-steel pins. Leave it for a few minutes to set in place, making sure that both ball races are flat. If you are making a multi-cylinder engine all the con-rods will then be precisely the same length. This is a very simple way of making a con-rod with ball races at both ends.

The con-rod on a model hot air engine takes only a small load, so there is no need to make them heavy. Some people make con rods of $1/4$in steel plate. Why do this when, for a model hot air engine, a bicycle wheel spoke will do? I also use stainless-steel welding rod.

CRANK

Most model engines have only a single crank disc, which is simply a disc with a crank pin in it and a hole for the main shaft in the centre. For the normal model hot air engine this is quite good enough.

For my boat engines twin-webbed cranks are used in an attempt to maintain a balance. It is not true balancing because I have just cut away the top sections to put a little bit of weight at one end to counterbalance the weight of the con-rod. Balancing is an art in itself and the average model engineer does not have the facilities to balance an engine properly, even in modern practice. Real balance is very rarely

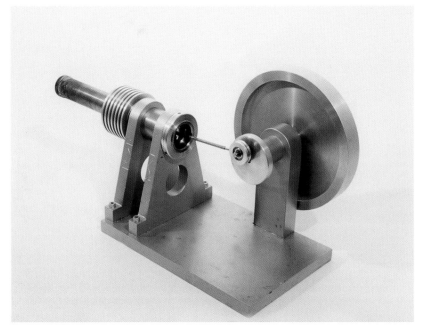

This lag engine shows how simple a crank disc and pin can be.

Not all crank discs have to be round.

Boat engine double crank.

achieved other than in multi-cylinder engines. In a car, with four- and six-cylinder engines, one component often counterbalances the other component. With single-cylinder engines there is always going to be a certain amount of imbalance, hence vibration. I make my crankshafts of aluminium if possible, as this way the revolving mass is kept to a minimum. Use a plain disc with a crankpin in, or, if using a shaft that requires bearings at both ends, a crankshaft that has two webs joined with a crankpin will be necessary.

If making a double crankshaft, initially drill the centre hole, turn the discs in the lathe and part them off from a bar. Then transfer the chuck that is still holding the bar to a lathe milling attachment and, ideally using a digital read-out for accuracy, mark the throw that will be equivalent to the stroke. Line up the milling jig on the centre line of the lathe and then use it to drill and ream the hole for the crankpin.

When using a double web crankshaft, make provision to clamp the crankpin. This allows it to be dismantled and enables the fitment of a

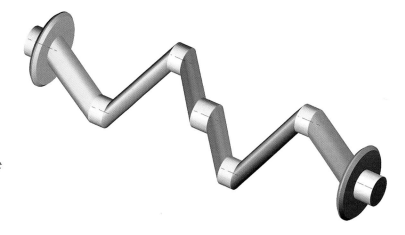

A diagram of the layout of an automotive four-cylinder in-line crank with balance weights removed for clarity.

ball race in the big end. At least one of the sides of the crankpin should be removable to enable the con-rod to slide on. One of the webs can then be tightened up. Part the two webs off, remove the waste material and then cross-drill one of them to take a nipping screw. Small Allen screws are suitable to nip the crankpin.

It is a very laborious task to make multi-webbed crankshafts from a solid bar. There would be a huge amount of waste. There would need to be a disc at either end with the journals being placed on a multiple-centres offset on a PCD, so that the crankpins can be turned for all the required throws. It is not a job to take on lightly.

One solution to this problem is to use large internal diameter ball races, which are reasonably light but with a very large centre hole. This way, the ball race becomes an eccentric and the main shaft becomes the offset, which is set at the stroke required. This makes a very good crankshaft. The disadvantage is that there is no direct counterbalance weight. This counterbalance weight could be attached to the main shaft to balance things up; also consider removing as much material around the ball race as possible.

For a twin cylinder they can be put at 180 degrees and one counterbalances the other; not completely of course, but it should reduce a lot of the imbalance. The big advantage of this type

A collection of ball races with a large internal diameter.

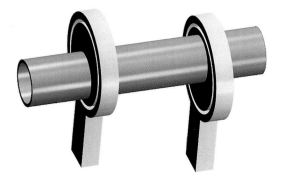

A simple method of achieving a crankshaft throw without having to machine an automotive-type crank.

139

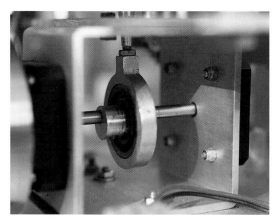

An example of a shaft using an eccentric bearing. (Reproduced with the kind permission of Graham Darlington)

of crankshaft is that it can have a plain shaft right the way through the middle, with appropriately offset con-rods and bearings at either end and a bearing in the middle if necessary.

If a four-throw crankshaft is required, all that is necessary is to lock the four eccentric ball races at 90 degrees to each other on the shaft to the distance to the centres of the cylinders (gudgeon pin). This makes a four-throw crankshaft with very little effort. It will run true because of the one central shaft all the way through. This is an ideal way of making a crankshaft, be it for two-, four-, six- or even eight-cylinder engines.

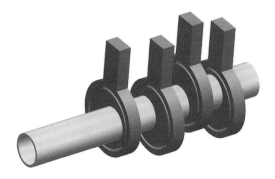

An example of a four-throw eccentric crank mounted in ball races.

I have made twin-webbed crankshafts with ball races on either side of the shaft so that the shaft is well supported. The crankpin is locked solid in one web; this way, the crankpin is taking the drive to the flywheel on the other end of the shaft. If there are a few microns of misalignment this will be taken up by the two bearings. Provided the locking screw remains tight, it will drive the other side of the crankshaft equally well and a free-running engine will result.

MAIN SHAFTS

Recently I have favoured twin flywheels, especially for boats, because the drive is taken from the OD of the flywheels.

Turn the seating for the ball race so that it gently slides in. Thread the end of the bearing housing and also machine the housing for the ball race either side at the other end, then part it off in the chuck.

Machining a bearing seating for a ball race.

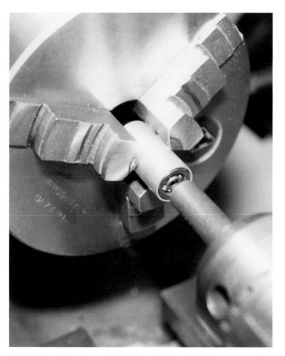

This bearing is pushed in with a drift mounted in the tail stock.

This shows the use of a bearing taken from a computer.

A shaft with a bevel.

In my early modelling days I used to press ball races in and then wonder why the ball race stopped working in a year or so, or made a terrible noise. The reason was because ball races compress and tighten the balls. To avoid this, place a smear of Loctite on to the inside rim of the bearing housing and gently slide the ball race in. This is a very delicate operation because if too much Loctite is used it may get into the bearings.

If the ID of the ball race is ¹/₄in, then a ¹/₄in diameter silver steel will not fit and should be lapped slightly for the ball race to fit. One advantage of utilizing bearing assemblies such as the shafts and bearings from computers is that the shafts have been ground to fit the ball races.

Ensure that the end of the shaft where it is parted off has the face at each end free of burrs. A little polishing is needed on the shaft. Place it in the three-jaw chuck and put a slight radius or chamfer bevelled on to the edge.

It is advisable to lap the silver steel down so that the ball race slides on in the area where the ball races are going to be required. Lap both ends of the shaft for about 1¹/₂in. It is surprising how little lapping is required to do this. A few minutes using a piece of wet and dry paper (worn wet and dry is fine) on the shaft when it is rotated in the chuck also achieves the same result. This polishes the shaft until the ball race goes on. It should be done evenly otherwise it won't be true. Ensure that the work is done from the end, keeping it parallel and only removing a small amount of material. This should be a straightforward operation.

Ball races should always be used for main bearings. It is okay to have plain bearings in an early or first model, but the difference made to the running of a hot air engine is considerable, so it is well worth using ball races. Spend the extra to get new ball races. Some ball races can be bought cheaply at shows, but they will be of an unknown quality and often have already been used and may be dusty, dirty or corroded. Names of the suppliers I use and recommend appear at the back of this book. They can supply ball races of very high quality to any size that may be required.

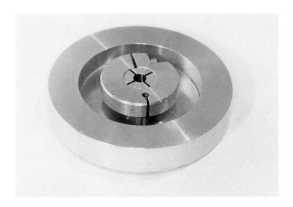

Stirling Silver II flywheel, collet and collet clamp.

The three bearings shown here are a standard bearing, a linear ball bearing and a bearing with a flange.

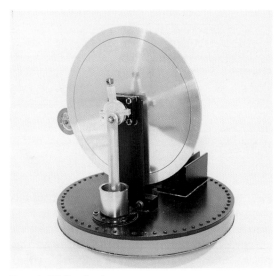

Low Δ T flywheel.

FLYWHEELS

There are various types of flywheels. They can be machined, fabricated or cast. Castings are available to buy, but overall I am not keen on using castings as they can be awkward to hold. A cast-iron flywheel would be ideal in weight, but on most of my engines they are made from aluminium, which seems to be quite satisfactory. An aluminium flywheel would be turned from a large-diameter bar or from pieces cut from heavy, large-walled tubing and have an aluminium disc let into it and bonded using Loctite.

A cast flywheel.

A pair of machined aluminium flywheels.

Over time, my flywheel construction methods have improved. One of my pet hates is to see any extraneous movement at all on a flywheel. It should run perfectly true. It is common to see a model running at a show that is beautifully made but the flywheel is wobbling away. I take a great deal of care in machining a flywheel. The real secret is to hold it securely and ensure that it is machined as true as possible.

Machining a Solid Aluminium Flywheel

Put a disc of aluminium in a large four-jaw chuck and skim the face. Ensure that the metal billet is tapped home against the chuck face with a soft-faced hammer while slowly tightening the jaws on the chuck. This helps to ensure that the flywheel is held as true as possible. Reverse the flywheel and machine to the required width, allowing a little extra material for the final finishing. As before, tap it home, ensuring that it is as flush as possible, and machine the face.

Remove the surplus material from the flywheel. The width of the outer rim should be decided and skimmed across the face.

Next, size the diameter of the boss.

Decide the wall thickness that will be left in the middle of the flywheel. Machine to a depth slightly under half of what the wall thickness will be from each side of the flywheel.

Mounting the flywheel in the chuck.

Sizing the outer rim.

Sizing the boss.

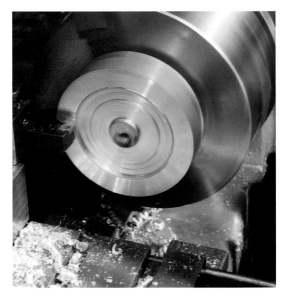

Machining a fine finish.

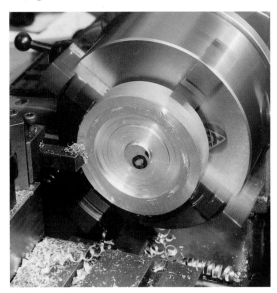

Facing the rim of the flywheel.

This tool is rounded or chamfered to ensure that it clears the internal flywheel radius when machining.

Machine the face of the outer rim and boss to ensure a good surface finish.

The diameter of the boss should be checked. Once satisfied with the boss and rim measurements, the cut should be squared off with a suitable tool that has a 'V' section or a radius underneath to clear the internal rim and the boss.

The cutting edges of the tool should be sharp and square. The work will chatter if the speed and the feed are too fast, so choose a slow speed using a fairly heavy cut to remove the surplus material.

A tool should be ground especially for this job. The nearest side of the tool needs to clear the inside radius of the flywheel. It needs a fair

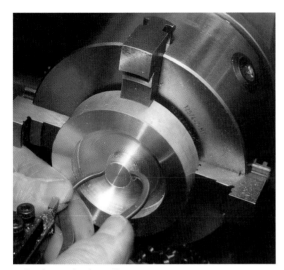

Checking the boss diameter.

Squaring up the cut.

as you go. When approaching the required depth, begin to take finer cuts to give a good finish. That means machining on all three surfaces in one operation – the outside diameter of the boss, the face and the inside diameter of the rim.

The final finishing.

Chamfering.

angle so that you don't scuff the metal that has already been turned.

Using digital read-outs on the lathe makes the job very easy because you do not need to stop constantly and measure the work. However, not everybody has this facility, in which case it is necessary to measure the depth

Remove the flywheel from the chuck and make sure it is clean. Reverse it, put it on the external jaws of the chuck and clamp it really tight. I have not found it necessary to put packing underneath.

The edge of the jaws have quite a large radius, so turn the key on the chuck tightly. Using a soft-faced hammer, make certain that the turned portion is hard against the chuck jaws, then really tighten and tap it home as before, otherwise you will get tool chatter. Always aim for the best machined surface possible on a flywheel on all faces.

The outer rim is now exposed, so take a cut across it to remove any irregularities and get it running true. Ensure that this has been achieved and make some fine cuts to achieve a good finish. This is the part of the flywheel and indeed the whole engine that will be on display most.

Centre-drilling the flywheel.

Finishing the outside of the rim.

Use a small centre drill and drill into the centre, then switch to a perfectly ground drill (I keep one specifically for this purpose). Go through the boss, ensuring that the drill is lubricated and that any swarf is cleared away frequently. Drill in and out progressively deeper a few times, each time removing the swarf.

Drilling the hole for the shaft.

The next task is to make a hole for the shaft, which is one of the most important jobs to get right on the flywheel. Centre-drill relative to the size of hole required.

Take your time with the feed to get this hole true and exactly through the middle, because the accuracy of all the other holes is dependent on the accuracy of this one. Open up the hole with increasingly larger drills until you reach slightly undersize (for example, 4.8mm for a 5mm hole).

Finally, open up the hole with a reamer, ensuring that the reamer is used to its full depth. As the reamer usually has a starting taper, you will get a very slightly tapered hole. Finish reaming by hand afterwards to achieve ultimate accuracy. Ream the hole right through, using the flywheel shaft as a gauge. There should be no slack and this must be accurately reamed if the flywheel is to run true.

Sizing the hole with a reaming tool.

The flywheel has now been faced, all of the surplus material has been removed, the rim has been turned and all that is left to do is to cross-drill to grip the shaft if using nipping grub screws. This should be done at an appropriate angle. The two holes should be indexed on the lathe, and drilled and tapped to hold Allen grub screws. This is done with a drilling attachment that fits to the cross slide and the chuck is locked, gripping the flywheel in place. The drill attachment drills at centre height in the lathe. Centre-drill to start and then drill the tapping-sized hole, drilling while the flywheel is still held in the chuck.

This gives an accurately placed hole. Index the chuck round 90 degrees and repeat the operation. If you are only attaching a small flywheel, then only one grub screw hole is required. I do the tapping by hand, as it is at an angle, and I use a long-handled tapping key in order to turn the tap, as it may be 2^1/2in to the centre of the rim.

An example of a cast flywheel.

A flywheel can also be made from a slice of a thick-walled aluminium or brass tube. Face both sides and counterbore them halfway plus half the width of the material to receive the 'flywheel wall' disc. Then glue this disc into place using Loctite.

Turn it round and put it into the chuck, using the external jaws to push outwards on the inside of the flywheel rim. Counterbore the inside of the tube again; this makes it look as though it is solid all the way up to the disc. Bore a hole for the boss and fix a boss bush blank in the centre with Loctite.

Drill as described in the previous section. This is a quick way of making a reasonable flywheel. I often use whatever material I happen to have at the time.

It is also possible to make a simple spoked flywheel by having an outer rim of brass or aluminium tube that has been cross-drilled. Drill the rim six times and the boss right through three times with an equal PCD division using an indexing tool, making a hole size corresponding to the spoke diameter. Leave the rim of the flywheel in the lathe chuck and mount the boss in the tailstock chuck to maintain true alignment. The boss bush blank should be left longer than the boss in order to allow sufficient material to hold it in the tailstock chuck. The boss should be lined up with the centre of the rim, and the spokes should be pushed through and bonded into place with Loctite. This is then left until completely cured. The boss blank should be parted off and drilled to fit on the shaft using the technique described for the solid flywheel.

Drill the hole in the centre of the boss as the last operation, making sure that everything is running true before drilling the hole. The flywheel can then be bonded to the shaft. Once dry, true it all up and skim it.

To make a flywheel using bicycle spokes, first make a boss. The end of the spoke will pass through a flange machined on to this boss. The

A simple flywheel using spokes.

spoke is bent through 90 degrees and the end of the spoke comes with one end already formed into a mushroom-shaped head that prevents the spoke from being pulled through the hole in the flange. The flange is positioned on the boss to take account of the diameter and the bend radius of the spoke. Therefore, the distance between the middle plane of the boss and the flange plane will be the distance from the flat underside of the mushroom head on the end of the spoke to the far edge of the material of the spoke after it has been bent through 90 degrees, minus half of the diameter of the spoke.

The flange should have a generous radius where it joins the outside of the boss. The 'axle' shaft of the flywheel should be bonded to the flywheel with Loctite. The rim of the flywheel should be made from a suitably heavy material. Drill the rim using an indexing tool according to the amount of spokes required (a minimum of twelve) and drill clearance holes to allow the spoke to pass through the rim with a minimal clearance.

Thread the end of the spoke to a length of a few millimetres more than the thickness of the rim and counterbore the rim to approximately half its thickness. Special nuts are required, which are drilled and tapped to screw on to the ends of the spokes. These are available from specialist bicycle parts suppliers. The OD of the nut should be a clearance fit in the counterbore and the diameter of the counterbore should be half the width of the rim or less. The length of the nut should be equivalent to the depth of the counterbore. The top should be slightly radiused and domed, and should have a thin slot machined across the top face. The slot will be used to screw the nuts on to the ends of the spokes as they are pushed through the clearance holes from the inside of the rim.

Hold the boss in the tailstock chuck of the lathe and hold the rim in the chuck. This will ensure that the flywheel boss will be as central to the rim as possible during the assembly of the components. Initially, hand-tighten the nuts on the end of the spokes through the flywheel.

Twin flywheels made using bicycle spoke technology.

Once all the nuts are fitted in this manner, tighten further, sequencing the tightening procedure by tightening the nut on the opposite side of the rim from the first nut to be tightened. Step and repeat this process using the maximum angle possible in the step rotation. Do not tighten up the nuts to their full tightness all in one go, but instead repeat the previous tightening process, slightly increasing the tension each time. It would be advantageous to use a torque wrench with an increase in torque settings for each step up to the maximum torque for an equivalent diameter bolt of the same material as the spoke.

To assist in maintaining the accuracy whilst tightening the nuts, spin the flywheel after each opposing pair of nuts has been tightened to see the effect it has had. Use a DTI (Dial Test Indicator) gauge to determine the eccentricity of the flywheel rim as well as the run out (wobble) on the side of the rim. The eccentricity can be slightly adjusted by the loosening and tightening of the opposing nuts. The whole structure may be made stronger by

angling the spokes so that they cross other spokes before reaching the hole in the rim. Crossing up to three spokes is acceptable, weaving them in and out of adjacent spokes. Once the assembly is true, place a drop of Loctite in the thread of the nuts and allow it to cure. The protruding ends of the spokes can then be trimmed and if desired a further aesthetic band may be used to cover the nuts on the outside of the rim.

On Low Δ T engines a perspex flywheel is all that is needed. First, make a disc of plywood and bolt it to the faceplate. Screw a square of perspex of the required thickness on to the plywood with wood screws, one in each corner. Drill a centre hole after you put it on the faceplate. Using a parting tool cut the disc out. Eye protection should be used when doing this.

These techniques are also used in the manufacture of a Low Δ T displacer piston, the only difference being the material used. The photographs show the displacer piston being made.

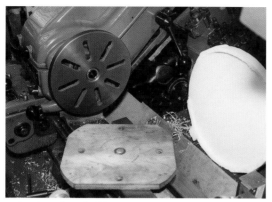

The faceplate holds the wooden backing plate to which the polystyrene is nailed.

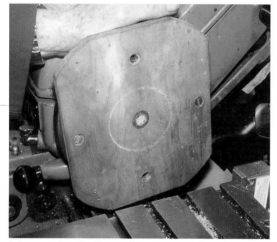

The wooden backing plate is screwed to the faceplate.

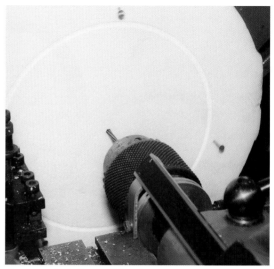

A centre hole is machined.

The disc is then cut.

The flywheel will probably have a flange on the shaft so that it runs true. Old Perspex may have warped with age and temperature. Turn the lathe by hand just before breaking through the Perspex, otherwise it may shatter. It should then just drop out.

The polystyrene for the displacer piston is nailed to the wooden backing plate.

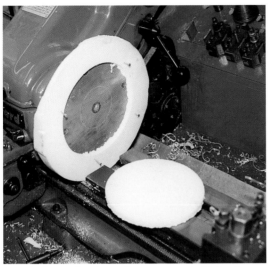

Ready for its final sanding, a Low Δ T displacer piston.

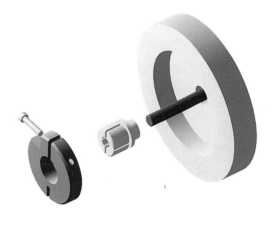

A collet with a clamp, flywheel and shaft.

I suggest using a nipping screw or some kind of cone collet clamp. If you are going to use nipping screws it is worth putting a flat on the shaft, either in the mill or with a file. When you tighten them down on the shaft you are then able to remove the flywheel without it getting stuck on the dimple created by the grub screw. Although the point of nipping it up is to stop the flywheel falling off, if it needs to be dismantled (and there usually are various reasons for this over time), it is worth taking care to remove the raised portion in a mill or with a file. You can also split the hub of a flywheel and put a clamp on the outside, or you can use a cone on the flywheel with a nut on the outside and the nut will just push the cone on.

I always try to find the simplest way to do things, but I never loose sight of ensuring that things run as true and as accurately as possible.

FRAME

Eventually, all of the items that have been made will have to be attached to a frame of some kind, such as a wooden base or an aluminium base with side frames screwed to it. This is dependent on what type of hot air engine is being made. If it is a model of an old hot air

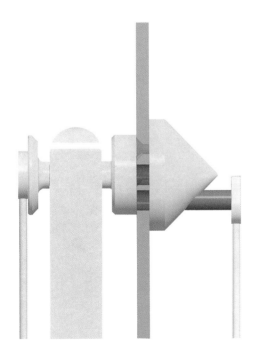

A view of a Low Δ T flywheel assembly.

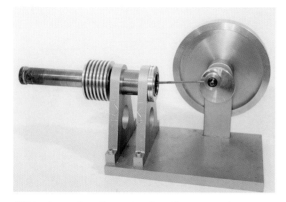

This shows just how simple a frame can be.

engine, the original shapes will be there to conform to and castings may well be used to attach all of these items together.

The copper water jacket of Stirling Silver II.

The hot and cold ends of the Stirling-Powered Buggy.

COOLING AND HEATING

If optimum performance is to be achieved, close attention should be paid to the heating and the cooling of an engine.

Cooling

WATER-COOLED
The most efficient way to cool the cold end is with a water jacket. To make a simple water

An aluminium water jacket.

jacket, make two holes, one for an inlet and another for the outlet, which will be fitted to tubes that will carry the water to a reservoir.

The configuration of the water jacket is such that it is a sliding fit over the displacer, and placed at the cold end. Turn a channel to accept an O-ring, then slide a brass tube or copper tube over the O-ring with a smear of lubricant. I sometimes add an internal ring to the brass or

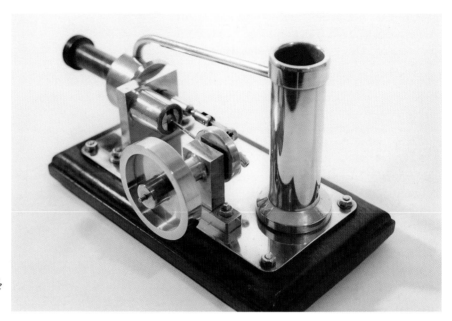

The cooling tower used on Gamma Mk III.

A centrifugal, engine-driven pump.

Stirling Silver I water jacket.

copper tube, which gives a larger diameter (and hence volume) water jacket (*see* Stirling Silver II). The advantage of doing it this way is that if any corrosion occurs after a few years, it is possible to dismantle the jacket easily and clean it up before reassembly.

A water-cooling tower can be made from a simple tube that mounts on the base. It should be large enough to contain enough water to keep an even temperature if using convection cooling. The most efficient and cost-effective choice of material would be to use aluminium or copper for its good thermal conductivity. Fins may be used, which will improve further on the efficiency of the cooling system, as will forced air, which will increase the relative cooling effect. So consider a fan and fins.

Pumped water is more efficient than convection, so make a small pump driven from the engine and pump the water round. In Sterling Silver I there is an electrically powered pumps, although it defeats the object of the engine a bit if you are having to supply electricity separately to pump the water.

When constructing the engine for Stirling Silver I, the water jacket and the block were milled out, which maximized the surface area that is cooled. A series of plates with gaskets in-between was used, which, when put together, form a water jacket with the same effect as the milled-out water jacket. This way, cooling water runs inside the block as well as the water jacket.

AIR-COOLED

Having lots of aluminium fins on the cold end increases the surface area and makes it more effective. Start with a block of aluminium and, using a parting tool, cut as deeply as possible, increasing the total area that is coming in contact with the air. In order to make the process as efficient as possible it is best to have as many thin fins as possible.

Parting off can be problematic. Some metals, especially aluminium, turn quite freely, but can jam and ruin a delicate fin. Cut to the depth required using a special Q-cut tool that has a purpose-made and shaped tip for parting in this type of circumstance.

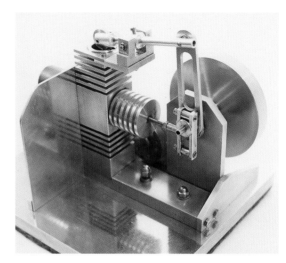

The cooling method on Gamma Mk V.

The aluminium chips turn inwards in the curl so that the tool doesn't jam in the cut as an ordinary parting tool would do. If you haven't the facilities for this, then an ordinary parting tool can be modified with a little groove in the middle cut with a very fine Dremel-type cutting or grinding disc. This gives a similar effect to the Q-cut tool.

The deeper the cut, the more important it is to ensure that a copious flow of lubricant is maintained. I use paraffin when cutting aluminium. The Q-cut parting tool I use is

A phosphor-bronze-lined power cylinder with cooling fins.

Cooling fins with an integral power piston and displacer.

Black cooling fins are more efficient.

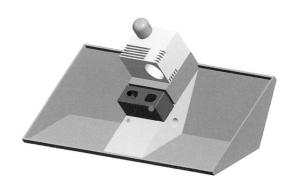

A proposal for the heat sink in a boat.

0.100in wide and cuts the corresponding groove size. In many of my engines, I have made the fins 0.200in apart with 0.100in slots. Cosmetically, it is better to have a thick fin at one end rather than a thin fin if your measurements don't work out precisely. You can also use, for example, a ¹/₁₆in parting tool that can be moved along 2 × 0.052in (0.104in) to get even thinner slots and fins. This will increase the surface area and thus the cooling.

Bore the finned heat exchanger so that it is a snug fit, making sure it remains in contact with

the cylinder in order to transfer the heat from the cold end to the fins. With a slitting saw and a milling machine it is possible to produce aesthetically pleasing fins on a square block, which also has the advantage of being easier to clamp.

Conclusion

In my future boats, blocks of aluminium will form part of the hull that is in contact with the water, perhaps with fins slotted on the underside. It makes sense to use the water to keep the cold end cool.

In Stirling Silver II the water is pumped through a cooling jacket. This water is picked up using a small tube very close to the propeller.

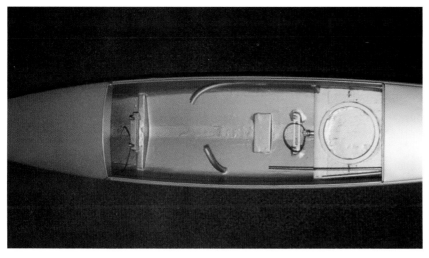

The two pipes shown in this top view are the cooling pipes. The long one is the intake and the short one is the water outlet.

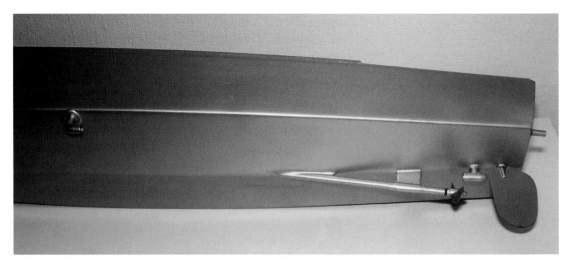

Shown from the underside, the intake can be seen near the propeller and the outlet is on the left of the photo.

This almost makes it an impeller, pushing water up the tube to the water jacket, where it is forced out from the underside of the boat. This is very efficient. Stirling Silver II runs for hours with the burner going full blast and the cold end remains quite cold. This is one of the attributes of Stirling Silver II's engine that has made it such a success. Although the engine capacity is not very large, it propels the boat very well and people are surprised at the power and speed of the boat. For making a small engine to run as a demonstration model the added complexity of water cooling is often unnecessary.

Heating

BURNERS

For a small hot air engine an appropriate burner is a methylated spirit burner. Many ordinary containers can be used to make a methylated burner, for example a small glass ink bottle. Ideally, these burners should be made of a non-ferrous metal. The basic requirements are a wick, an easy method of filling and a vent hole.

Safety must come first before anything else. All my burners have a little air vent in the

A selection of methylated spirit burners.

burner top so that as the methylated spirit is used it can be replaced by air in the tank. Some burners and some articles I have seen on burners often show no vent at all, relying completely on the wick to be a reasonably loose fit in the tube that supports the wick, to let air go in. This is an unsatisfactory design and a vent should always be provided. This may be a small brass or steel tube, silver- or soft-soldered to the top of the burner, which takes the vapour away from the flame.

If a burner has an air vent too close to the wick, the methylated spirit vaporizes inside the burner and it lights with two flames. This may

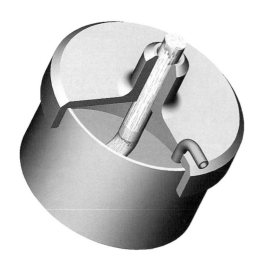

Cutaway view of a spirit burner showing the wick and breather tube.

A spirit burner with a coil of wick material.

cause an explosion because you are burning the gaseous mixture of the methylated spirit and it heats the burner up even more without allowing any air in.

A few years ago, I experienced a methylated burner where the vent hole was blocked. It was a steel burner and some dirt had blocked the vent hole. The methylated spirit started to boil, the burner blew and there was a mass of very hot methylated spirit thrown into the air. Looking at the bench it was on there was a metre square alight. There was a large container of water on the floor, so I picked this up and dowsed the bench.

The general advice of the Fire and Rescue Services is that an appropriate fire extinguisher should always be readily available when using flammable liquids. In this case a simple bucket of water was sufficient but would not be appropriate for an oil fire, so ensure that the extinguisher is suitable for the purpose. You should never attempt to tackle a fire if it puts you in danger and smoke alarms should always be fitted. If in doubt, get out, call the Fire Service and stay out.

Make sure that the vent of the burner is not blocked and check it each time before use. Keep a tool that is suitable for clearing or cleaning the vent hole handy, or even attached to the burner.

Wicks for paraffin lamps can be purchased from ironmongers. Fibre-glass types of wick are also available. Some model railway societies use fibre-glass strands that are twisted together and pushed into the main burner hole to form a wick. The advantage of this type of wick is that it does not burn and the capillary action draws the fuel up when it is tightly packed. I have used this method, but it can be tricky to get in the burner hole.

Methylated spirit can often be of very poor quality with a lot of water and dye in it. When using this type of methylated spirit and if your burners are not used daily, it can be a little tricky to get the burners alight again. Because the water rises to the top of the wick it has to be dried with a piece of rag before you can light it.

PROPANE OR BUTANE GAS

It is easy to make a little tank by silver-soldering or brazing two ends into a piece of tube. A special filler is needed, which can be purchased at shows or via mail order companies. I use the gas cylinders that are commonly used for

A pressurized gas tank in the Stirling-Powered Buggy.

An unmounted pressurized gas tank.

jet ends is where the air opening begins. If a $3/8$in diameter mixing tube is used, it should be matched with a $3/16$in square hole or four cross-drilled holes that would give you the flow of air to mix with the gas. To make the flame end of the burner, find a small round pot made from an appropriate metal and fill it with a cleaned stainless-steel scouring pad or kettle defurring material. Break some off, roll it into a ball and ram it into the end of the tube rather like a 'smoking pipe'.

A collection of gas burners.

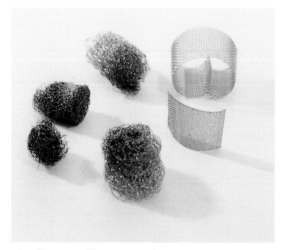

A collection of burner meshes.

camping. My smaller individual models have their own gas tank that can be bought complete with a jet, a valve and the filler. It is possible to make your own, but in my opinion and from a safety point of view it is worthwhile spending a little extra to purchase a tank which comes with a safety test certificate and some peace of mind.

The gas burners themselves can come in a wide variety of shapes and sizes – ring, square and semicircular tubes. Often they are similar to a Bunsen burner. For small burners, a number three jet is recommended. Where the

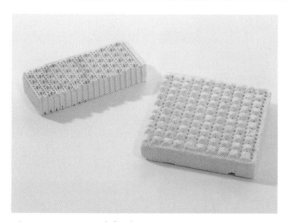

Ceramic material for burners.

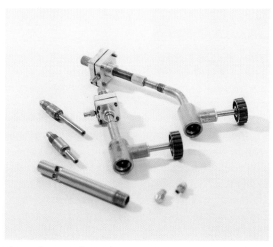

Jets, pipes and regulators.

Using this type of material the stainless-steel mesh acts as a flame trap. The flame cannot return back to the mixing tube and it also preheats the air and the gas as it comes through so you get a lovely blue flame.

Ceramic burner material is available and some suppliers make ready-made burners, as well as jets and gas tanks. If using ceramic it needs to be cut to shape and fitted to the box or tube it will be mounted in. It crumbles very easily, so care with cutting and storage is required.

The ceramic material creates incandescent heat and glows red, giving a radiant heat. A small box with lots of tiny holes drilled in it is in my opinion inferior because it is difficult to achieve equal flame heights.

WORKSHOP LAYOUT AND PLANNING

When planning to build or reorganize a workshop, its position, size and entrance will dictate many other characteristics of the layout of the workshop. If a workshop has already been built, or a space is available to convert to a workshop, then this will involve making the best of what is available. The ideal workshop is as large as possible and with plenty of light. A north-facing window gives the best light without the glare of the sun. Most model engineers build up their workshop over a period of many years, gradually obtaining more tools and equipment. As time goes by more and more space is needed. It is a good idea to look ahead and if possible start with as large a workshop as you can.

Before designing the workshop it is useful to make a scale model or a plan view to determine where each machine will go and ensure there is enough space around the machines to work unhindered. Both a milling machine and lathe require space to work at the front and at both sides. Many model engineers' circumstances dictate that they have to work in a confined space. It is possible, although not ideal, to operate a milling machine on a bench and a lot of model engines can be made with just a drilling machine, so this should be the first priority if finances don't yet permit the purchase of a lathe or milling machine. A mechanical hacksaw makes life a lot easier and the sooner one is acquired the more engineering can be done without wearing out your arms.

The position of a lathe should take priority as this will be the item most used and the ideal position is next to a north-facing window. My own workshop was already built with the house and originally the lathe was placed on a north-facing wall that does not have a window.

Myford Super 7 lathe and tools.

Eventually I moved it to a south-facing wall where there is a full-length window and plenty of light, but in the winter the sun is so low that it is necessary to draw the blinds and put the lights on.

It is a good idea to have some empty bench space next to the lathe so that when working you have space to put tools down ready to be tidied away once the job is finished. It is also important to have space to sit down and think when a problem occurs, better still with the project on the bench in front of you, so having empty space is as important as having places for machines.

The floor of my workshop gets a new coat of paint every year using garage floor paint. Light grey makes it easy to see when something has been dropped. It is easy to sweep round with a soft broom after each job, then vacuum to ensure the floor is clean. A dirty or untidy floor could cause you to trip or fall and safety in the workshop should always be the first priority.

Keep tools in the areas in which they will be used. The lathe has a large variety of attachments, such as tool holders, drills and chucks, and I keep these in a tool rack immediately behind the lathe. The main function of this is availability. You can put your

hand out, pick up a tool, use it and put it back. This not only saves time in finding the tools, but means they don't have to be cleared up at the end of the day.

Good lighting is very important. Daylight fluorescent strip lights are the best option; they should be as long as is practical. In addition to this each machine should have small portable lights that can be shone directly on to the work. If using a lot of rotating machinery it is important to consider that rotating parts, if running at a multiple of 50Hz (cycles per second), the same frequency as mains electricity, can under certain conditions appear to be stationary. Fluorescent lamps are available that work on a different or offset phase. If you have the luxury of a three-phase supply, then ensure that at least one light is wired on a different phase.

Shelving is also very important and almost everything I have is stored in boxes on the racks and shelves made to take them – bearings, hot ends, glass tubes, machined parts from previous experiments, pencils, and so on. A whole range of screw sizes are kept in containers, as are nuts ranging from 10BA to 6mm. Every box and container is labelled, so all that is required is to look quickly along the shelves for the appropriate container.

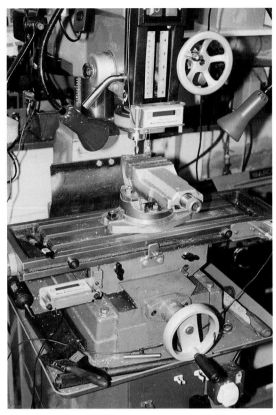

Warco Minor milling machine.

Aluminium stock.

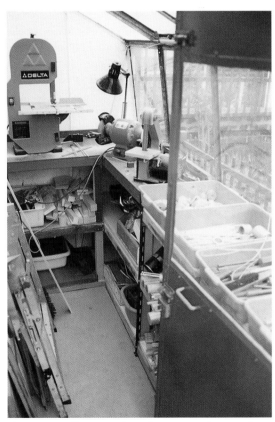

Dirty area for operations that create dust, such as grinding, cutting, sanding and linishing, which ideally should be kept separate from areas used for precision work or measuring.

To store taps and dies I use small plastic drawers that can be bought at DIY stores. I use a separate drawer for each size. For example, a drawer labelled 0BA has all taps, dies, a tapping drill and a clearance drill for 0BA inside. This is repeated throughout the range so that when a particular size is needed everything is available in the one container and this eliminates the need to stop and look up the thread chart to see which tapping drill or clearance drill is needed for that size. These are never left lying around, but always returned to the correct container immediately after use. The same system is used for reamers.

The purpose of such tidiness is to save time. It makes it possible to go into the workshop knowing that the right tools are to hand, get a piece of metal and make four completed bushed parts within twenty minutes.

Warco mini lathe in my indoor workshop.

People visiting my workshop often remark how tidy and well laid out it is compared to their own messy workshop. Sometimes, like everyone, I get untidy. I keep putting things down on the bench until suddenly I can't find something amongst the clutter of tools. That is the time to stop and put everything away before continuing with the work. Always try to put drills and reamers straight back into their containers as it wastes a lot of time to have to check each one with calipers to see what size it is.

Some people make their own tools, for example jigs and fixtures and cutter grinders, or stop to make a reamer from a piece of silver steel. I used to do this, but now instead of grinding a cutter or sharpening a drill I prefer to use the time on a project and buy new tools when they wear out.

I have recently installed an indoor workshop with an assembly bench, mini lathe on its own stand, milling machine and under-shelf lighting to throw light on to the bench. There are also flexible lights to place light wherever it is needed. Behind the lathe there is a rack made of wood with a range of drills from 1mm to 6mm. They are spaced further apart than in the containers they come in, making it easier to pick up the required drill, use it and put it back. If a drill is left on the bench, then it is easy to see from the empty hole in the rack where to put it back.

I constantly change and move things when I decide they would be better placed elsewhere. I spent many years laying out and rearranging factory floors, so laying out my own workshop comes naturally. It is amazing how much extra space can be made with the right layout. The aim is always to design the workshop for maximum convenience and safety.

Majic Mill in my indoor workshop.

17 Plans – Replica Stirling Silver II (Single-Cylinder Beta Engine)

This chapter is devoted to the package of engineering drawings that should enable a competent engineer to make a version of the Beta engine that powers Stirling Silver II. The drawings are drawn to a professional engineering standard, which gives a model engineer the choice as to how far they wish to maintain the tolerance levels specified. Every model engineer will have different capabilities and equipment, and the drawings take account of this. This engine is designed to be reasonably straightforward to build and to give an insight into making a powerful model Stirling engine that can be adapted for use in solar, model automotive and marine applications. The finished engine should give many enjoyable hours of operation and will be an achievement of which you can be justifiably proud.

- There are various standards that have been used in compiling these engineering drawings. The standards used for draughting are BS 308 parts 1, 2 and 3. The standard for limits and fits is BS 4500. The material designations are typically in accordance with the International Standards Organization (ISO). For steel the European Norm designation has been used. The ISO designations are used for heat treatment data. The primary unit of measurement is millimetres and some care has been taken to specify imperial measurements where appropriate, albeit in millimetres (a $^1/4$in shaft will be specified as 6.35mm). Generally, an alternative dimension will be given if close

enough to an imperial size, for example: 12.70 ($^1/2$in).

- The general dimensioning tolerances are specified (unless otherwise stated) by the number of decimal places or zeros on the dimension, for example:

  1 = ± 0.5mm tolerance
  1.0 = ± 0.1mm tolerance
  1.00 = ± 0.05mm tolerance
  1.000 = ± 0.025mm tolerance.

- Angular tolerances are specified thus:

  0.0 = ± 0.5 deg.
  0.00 = ± 0.1 deg.

- Multiple occurrences of the same feature are specified:

  TYP (for many occurrences), or
  TYP 4 (for four occurrences).

- Dimensions that are shown thus:

  (12.43) as opposed to 12.43 means that this dimension is the dimension that is the primary dimension of the component and every care should be taken to make this as accurate as possible.

- Where a dimension is shown with the word 'Ref' afterwards, it indicates that this dimension is for reference only and should not be used as a definitive measurement, but is included as a courtesy often to indicate material sizes. Generally it is bad practice to dimension the same part of a component more than once as this leads to confusion and inaccuracy and where this has occurred it has been to assist in the clarity of the drawing.

- When indicating a diameter it is shown thus:

  a 5mm diameter will appear as Ø 5.0
  a 2.5mm radius will be shown as R2.50.

- Holes linked geometrically that are symmetrical in series or controlled by the same geometric tolerancing method are usually connected with a centre line.

- Hidden lines are shown as a feint regular dash. Centre lines are shown with a long and a short alternating dash and are slightly less feint.

- The metric system is used for threads, as wherever possible the actual tapping size hole has been specified along with the 'M' series designation and the pitch of the thread, for example: Drill Ø 2.5 Tap M3 × 0.5. The abbreviation used to indicate a countersunk hole is C'sk followed by an angle designation which is to determine the angle on the cutter, for example: Drill Ø 2.5 Tap M3 × 0.5 C'sk 90. Usually the depth of the countersink would be specified. Because the potential readers of this book will have access to a wide variety of fasteners, these should be measured and then countersunk to the appropriate depth and angle.

- Where geometric tolerances have been used, these are in accordance with BS 308 part 3 and an appendix has been provided (see Appendix VI) to help understand what it means. The prime function of BS 308 part 3 is to control the accuracy of the form, the direction, the location and running tolerances from a specified datum.

- For clarity, threaded parts have been specified rather than drawn.

- Where ream is specified the holes should be drilled undersize and then reamed to the appropriate size.

- On the material designation heat treatment on these drawings, H22 may be replaced by H4. For all heat treatments the relevant standard for the material should be consulted and your metal supplier may well have good information and be able to tell you what he stocks.

- Make sure that careful attention is paid to the notes on each drawing, as they differ and may contain information not specified elsewhere. Some drawings have a hole chart included and here the corresponding letter in the chart has been placed as close as is reasonably possible to the hole it references. The number of items required is specified on the drawing and in the flow chart.

- The parent and child information on each drawing indicates which components a particular item is fitted to and which components fit to the item in question. This information is also contained on the flow charts and, simply by following the arrows, it is possible to determine in which order the parts fit together. When referencing data on the flow charts, to identify what fasteners are required to fit each component to its parent, look in the top third of the flow chart box, where it also shows the number of fasteners required in brackets. Items in the flow chart that are underlined and where the flow chart box is bounded by a double line indicate that no child parts are fitted beyond this component.

- The scale on the drawings has been deliberately omitted because of the inaccuracies of the printing process. It is bad practice to measure from drawings.

- Cross-sectional hatching is shown in accordance with BS 308 and ANSI 31.

- The general assembly and subassemblies show parts that are identified by their drawing number. Drawing numbers that start BAA5 indicate subassemblies and ones that start BAA0 indicate detail drawings of individual parts. Where possible, fastener information has been included on the assembly level drawings, as has any bonding or joining information. The designation for the bonding system is primarily using Loctite engineering adhesives. Should you choose to use different adhesives, you should ensure that they match the properties of the adhesives specified. Bonding information is shown on subassemblies with a dotted line to an arrow. Large parts are shown with a solid line to a dot unless required to indicate a component that was too small to be indicated satisfactorily with a dot. Where space does not permit the full title of the joining technique to be displayed, Loctite 641 is indicated with L6 and Loctite 248 is indicated with L2.

- For a graphical representation of BS 308, as applied to the engineering drawings in this book, refer to Appendix VI.

- A spacer may be required to take account of tolerance and machining differences. This spacer, if required, is fitted between the power cylinder (BAA009) and the front of the crankcase (BAA018). It is just a simple aluminium ring which ensures that the displacer piston (BAA046) does not contact the end of the displacer cylinder (BAA0050).

DRAWING UPDATES

To ensure the latest updates to drawings are used please check www.stirlingengineer.com using BAA as the project code.

Latest Updates
BAA027 , BAA030, BAA031, BAA048.

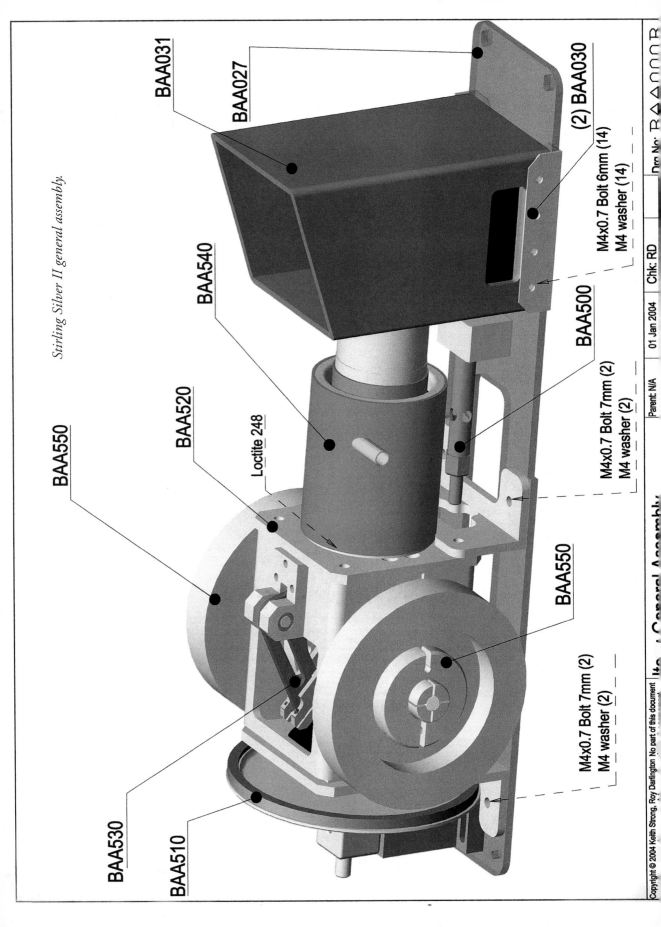

Stirling Silver II general assembly.

BAA031

BAA027

(2) BAA030

M4x0.7 Bolt 6mm (14)
M4 washer (14)

BAA540

BAA500

BAA520

Loctite 248

M4x0.7 Bolt 7mm (2)
M4 washer (2)

BAA550

BAA550

BAA530

BAA510

M4x0.7 Bolt 7mm (2)
M4 washer (2)

Copyright © 2004 Keith Strong, Roy Darlington No part of this document

Parent: N/A   01 Jan 2004   Chk: RD

Drg No: R△△□□□R

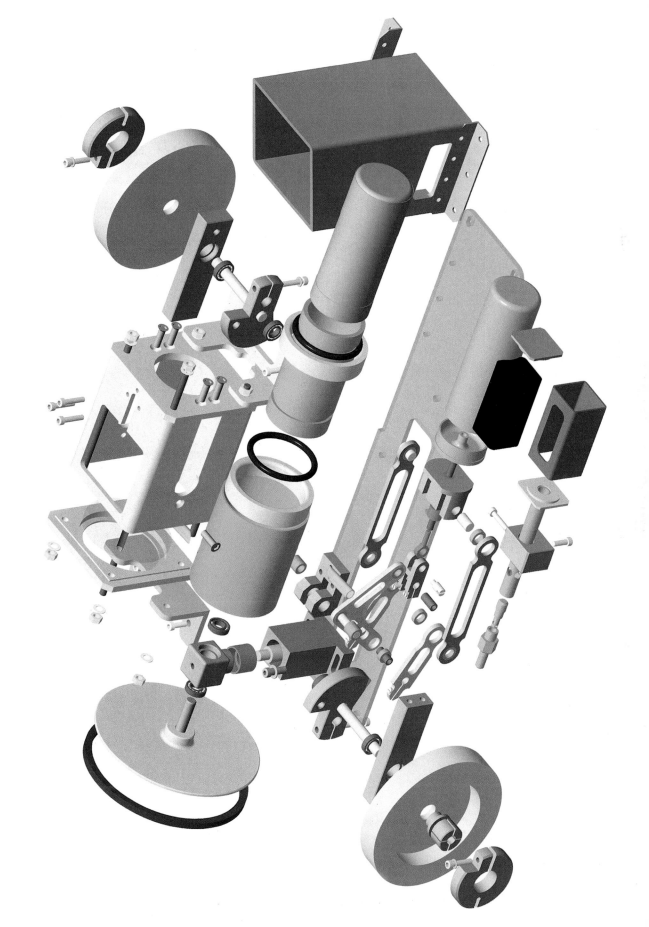

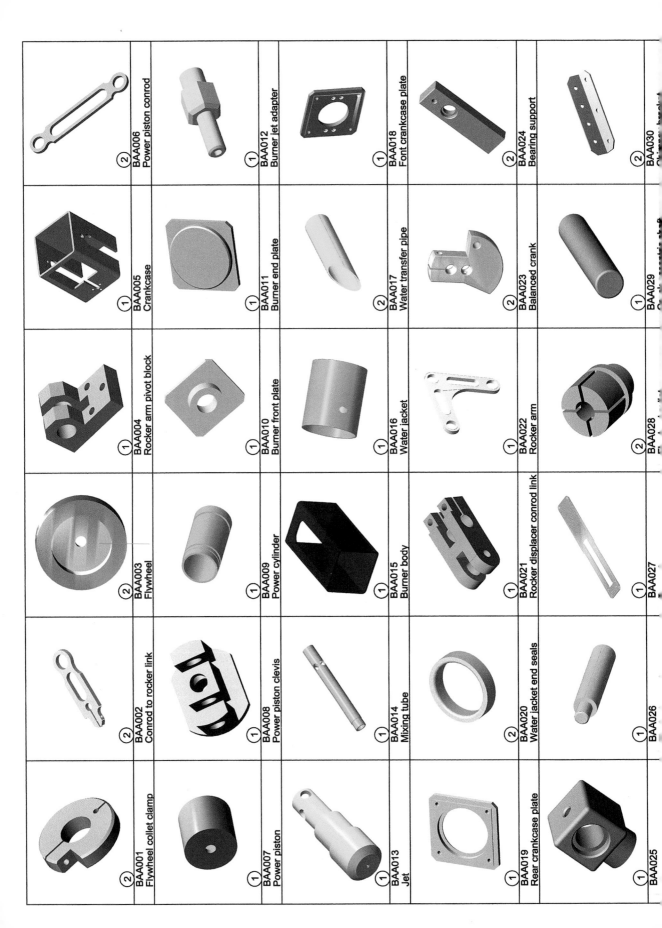

② BAA001 Flywheel collet clamp	② BAA002 Conrod to rocker link	② BAA003 Flywheel	② BAA004 Rocker arm pivot block	① BAA005 Crankcase	② BAA006 Power piston conrod
① BAA007 Power piston	① BAA008 Power piston clevis	① BAA009 Power cylinder	① BAA010 Burner front plate	① BAA011 Burner end plate	① BAA012 Burner jet adapter
① BAA013 Jet	① BAA014 Mixing tube	① BAA015 Burner body	① BAA016 Water jacket	② BAA017 Water transfer pipe	① BAA018 Font crankcase plate
① BAA019 Rear crankcase plate	② BAA020 Water jacket end seals	① BAA021 Rocker displacer conrod link	① BAA022 Rocker arm	② BAA023 Balanced crank	② BAA024 Bearing support
① BAA025	① BAA026	① BAA027	② BAA028	① BAA029	② BAA030

Part No.	Description	Qty
BAA031	Chimney	1
BAA032	Rear crankcase bracket	1
BAA033	Front crankcase bracket	1
BAA034	Crankcase inspection window	1
BAA035	Rocker pivot bush	2
BAA036	Rocker middle pivot shaft	1
BAA037	Rocker arm bush	3
BAA038	Rocker end shaft	2
BAA039	Power piston conrod bush	6
BAA040	Kettle defurrer	1
BAA041	Displacer big end	1
BAA042	Link shoulder screw	1
BAA043	Displacer shaft	1
BAA044	Crankshaft	2
BAA045	Displacer piston end plug	1
BAA046	Displacer piston	1
BAA047	Power piston displ. conrod bush	1
BAA048	Burner tube support	1
BAA049	Power piston clevis shaft	2
BAA050	Displacer cylinder	1
BAA051	O-Ring	2
BAA052	Drive post	1
BAA053	Drive post bush	1
BAA054	Timken S614C or MCK	6
BAA055	Final drive wheel	1
BAA056	Oil filter rubber ring	1
BAA057	Final drive shaft	1

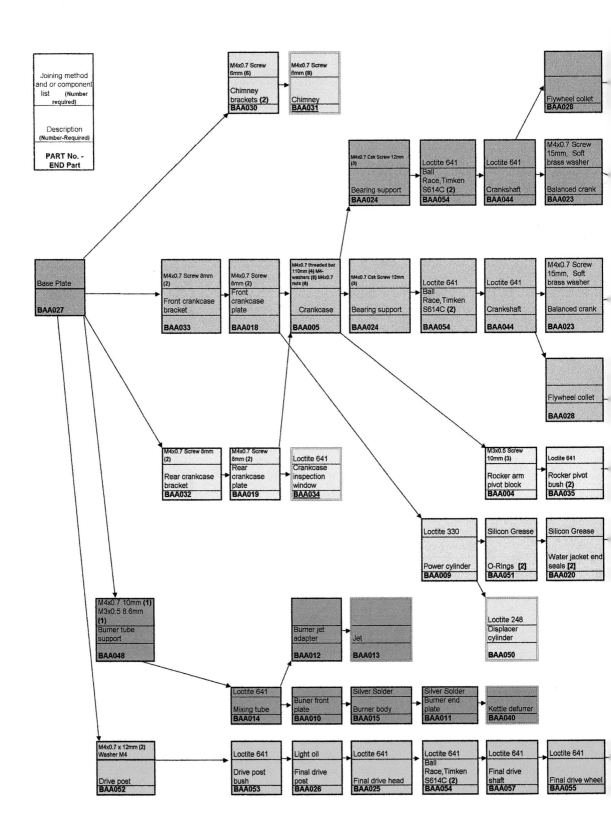

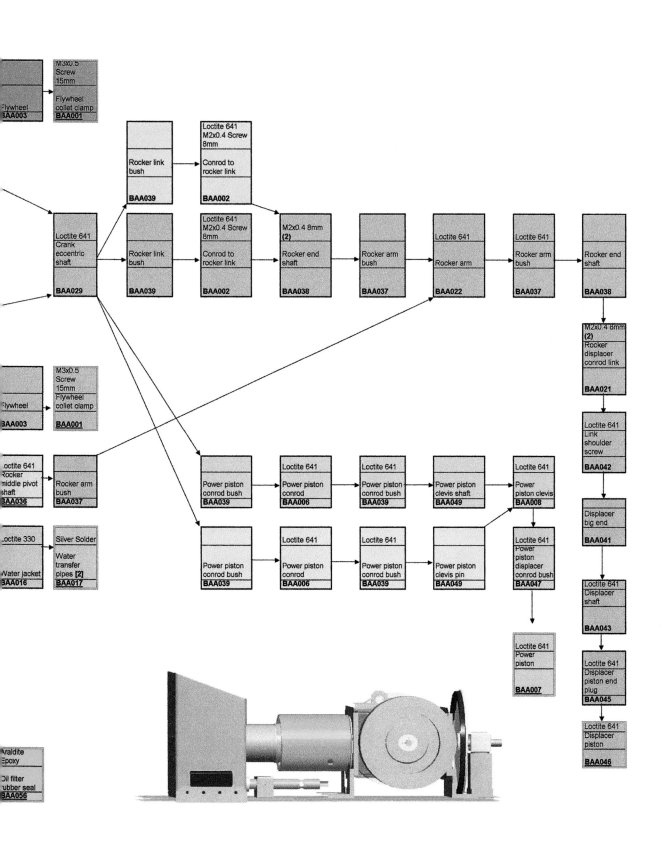

M3x0.5 Screw 15mm

Flywheel collet clamp
BAA001

Flywheel
BAA003

Rocker link bush
BAA039

Loctite 641 M2x0.4 Screw 8mm

Conrod to rocker link
BAA002

Loctite 641 Crank eccentric shaft
BAA029

Rocker link bush
BAA039

Loctite 641 M2x0.4 Screw 8mm

Conrod to rocker link
BAA002

M2x0.4 8mm (2)

Rocker end shaft
BAA038

Rocker arm bush
BAA037

Loctite 641

Rocker arm
BAA022

Loctite 641

Rocker arm bush
BAA037

Loctite 641

Rocker end shaft
BAA038

M2x0.4 8mm (2)
Rocker displacer conrod link
BAA021

Loctite 641 Link shoulder screw
BAA042

M3x0.5 Screw 15mm

Flywheel collet clamp
BAA001

Flywheel
BAA003

Displacer big end
BAA041

Loctite 641 Rocker middle pivot shaft
BAA036

Rocker arm bush
BAA037

Power piston conrod bush
BAA039

Loctite 641

Power piston conrod
BAA006

Loctite 641

Power piston conrod bush
BAA039

Power piston clevis shaft
BAA049

Loctite 641

Power piston clevis
BAA008

Loctite 641 Displacer shaft
BAA043

Loctite 330

Water jacket
BAA016

Silver Solder

Water transfer pipes [2]
BAA017

Power piston conrod bush
BAA039

Loctite 641

Power piston conrod
BAA006

Loctite 641

Power piston conrod bush
BAA039

Power piston clevis pin
BAA049

Loctite 641 Power piston displacer conrod bush
BAA047

Loctite 641 Displacer piston end plug
BAA045

Loctite 641 Power piston
BAA007

Loctite 641 Displacer piston
BAA046

Araldite Epoxy

Oil filter rubber seal
BAA056

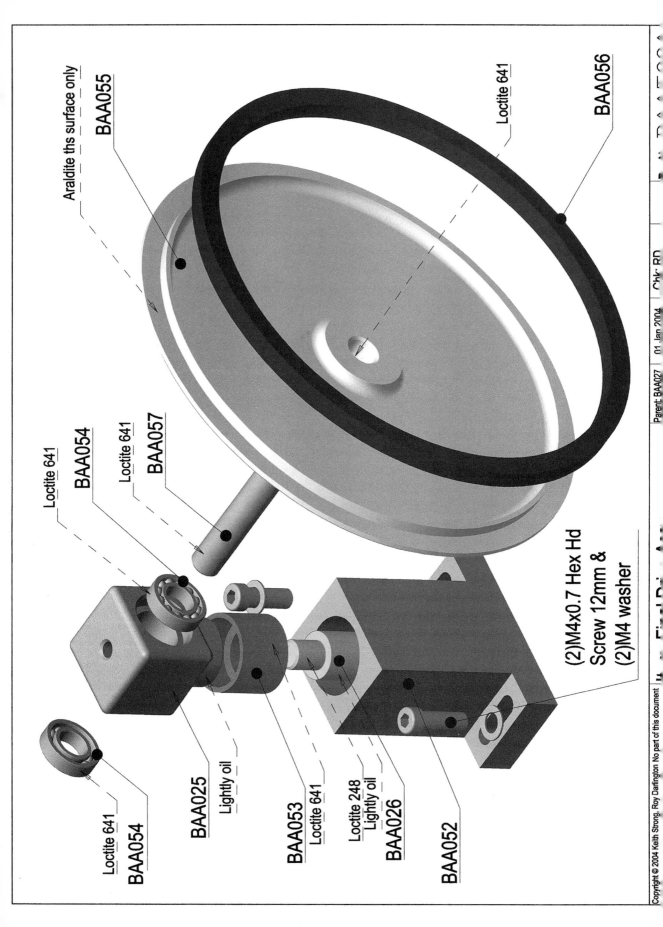

Araldite ths surface only

BAA055

Loctite 641

BAA056

Loctite 641

BAA054

Loctite 641

BAA057

Loctite 641

BAA054

BAA025

Lightly oil

BAA053

Loctite 641

Lightly oil

Loctite 248

BAA026

BAA052

(2)M4x0.7 Hex Hd
Screw 12mm &
(2)M4 washer

Copyright © 2004 Keith Strong, Roy Darlington No part of this document

Parent BAA027   01 Jan 2004   Chk: RD

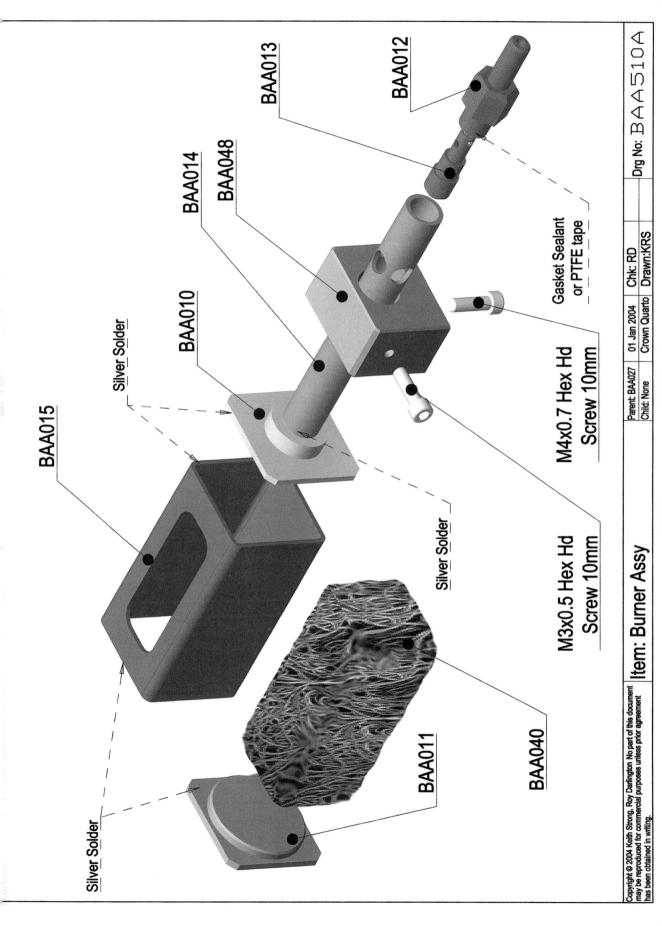

BAA013

BAA012

BAA014

BAA048

Gasket Sealant
or PTFE tape

BAA010

Silver Solder

BAA015

M4x0.7 Hex Hd
Screw 10mm

Silver Solder

M3x0.5 Hex Hd
Screw 10mm

BAA011

BAA040

Silver Solder

Copyright © 2004 Keith Strong, Roy Darlington No part of this document may be reproduced for commercial purposes unless prior agreement has been obtained in writing.

Item: Burner Assy

| Parent: BAA027 | 01 Jan 2004 | Chk: RD | Drg No: BAA510A |
| Child: None | Crown Quarto | Drawn:KRS | |

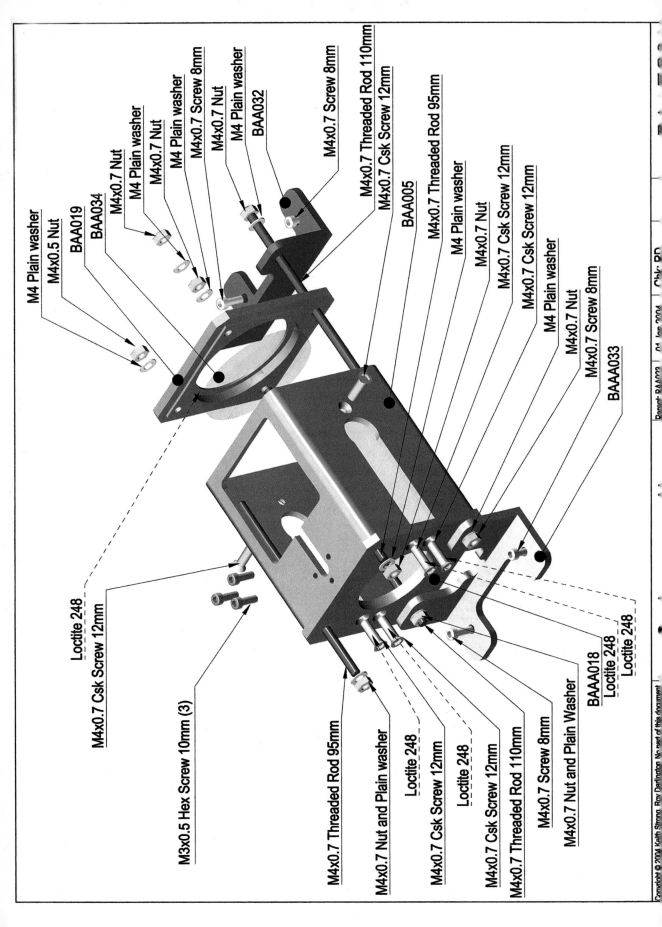

M4 Plain washer

M4x0.5 Nut

BAA019

BAA034

M4x0.7 Nut

M4 Plain washer

M4x0.7 Nut

M4 Plain washer

M4x0.7 Screw 8mm

M4x0.7 Nut

M4 Plain washer

BAA032

M4x0.7 Screw 8mm

M4x0.7 Threaded Rod 110mm

M4x0.7 Csk Screw 12mm

BAA005

M4x0.7 Threaded Rod 95mm

M4 Plain washer

M4x0.7 Nut

M4x0.7 Csk Screw 12mm

M4x0.7 Csk Screw 12mm

M4 Plain washer

M4x0.7 Nut

M4x0.7 Screw 8mm

BAAA033

Loctite 248

M4x0.7 Csk Screw 12mm

M3x0.5 Hex Screw 10mm (3)

M4x0.7 Threaded Rod 95mm

M4x0.7 Nut and Plain washer

Loctite 248

M4x0.7 Csk Screw 12mm

Loctite 248

M4x0.7 Csk Screw 12mm

M4x0.7 Threaded Rod 110mm

M4x0.7 Screw 8mm

M4x0.7 Nut and Plain Washer

BAAA018

Loctite 248

Loctite 248

Copyright © 2004 Keith Strong, Rby Darlington No part of this document

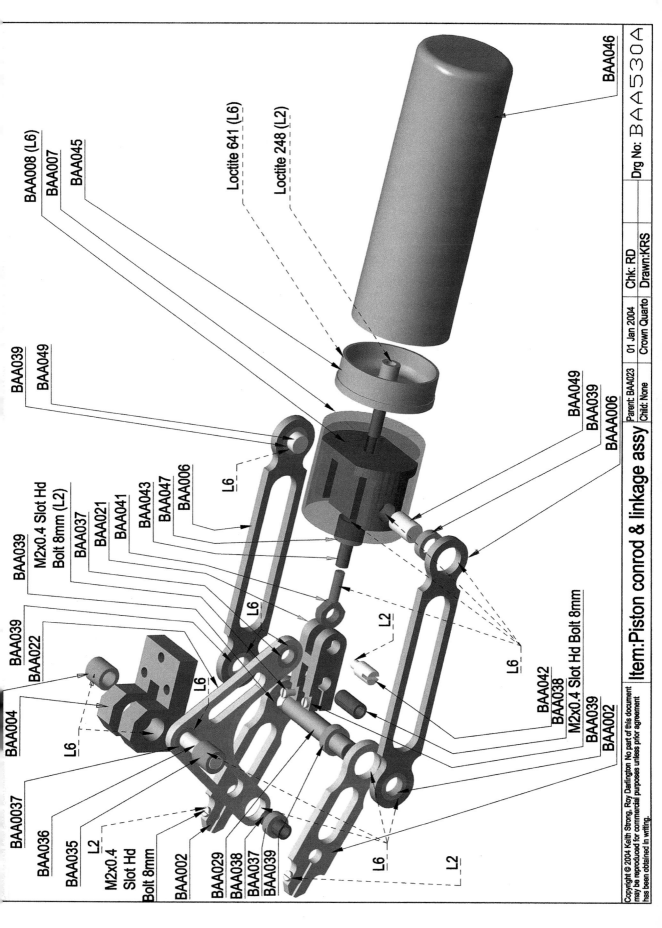

BAA008 (L6)
BAA007
BAA045

Loctite 641 (L6)
Loctite 248 (L2)

BAA039
BAA049

BAA039
M2x0.4 Slot Hd
Bolt 8mm (L2)
BAA037
BAA021
BAA041
BAA043
BAA047
BAA006

L6

BAA049
BAA039
BAAA006

BAA039
BAA022
BAA004

L6

BAA037
BAA036
BAA035
L2
M2x0.4
Slot Hd
Bolt 8mm

BAA002
BAA029
BAA038
BAA037
BAA039

L6
L2

L6
BAA042
BAA038
M2x0.4 Slot Hd Bolt 8mm
BAA039
BAA002

L2

BAA046

Drg No: BAA530A

Copyright © 2004 Keith Strong, Roy Darlington No part of this document may be reproduced for commercial purposes unless prior agreement has been obtained in writing.

Item:Piston conrod & linkage assy

Parent BAA023
Child: None

01 Jan 2004
Crown Quarto

Chk: RD
Drawn:KRS

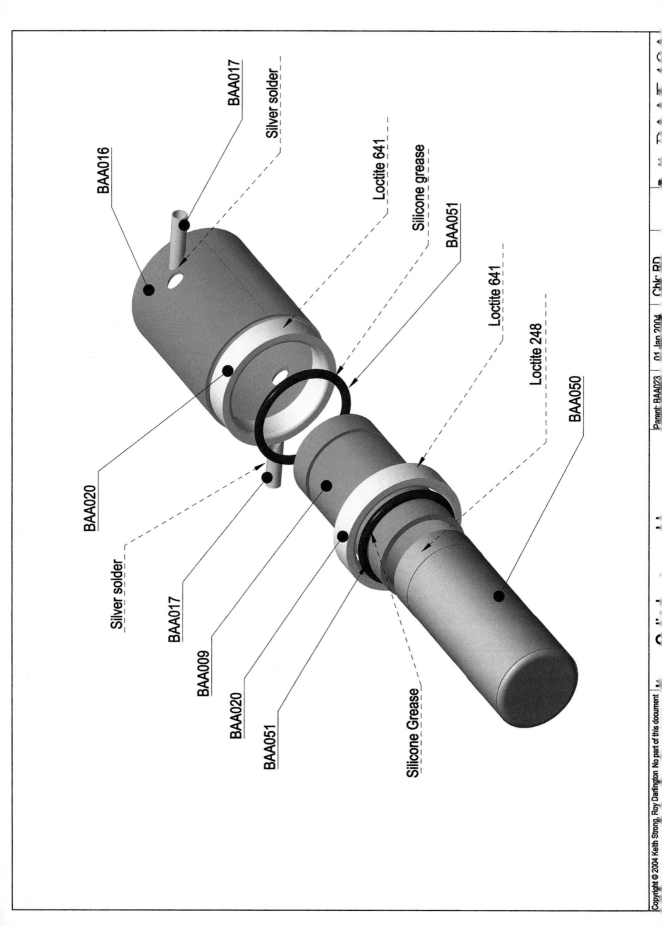

BAA016

BAA017

Silver solder

Loctite 641

Silicone grease

BAA051

Loctite 641

Loctite 248

BAA050

BAA020

Silver solder

BAA017

BAA009

BAA020

BAA051

Silicone Grease

Copyright © 2004 Keith Strong, Roy Darlington No part of this document

Parent: BAA023 | 01 Jan 2004 | Chk: RD

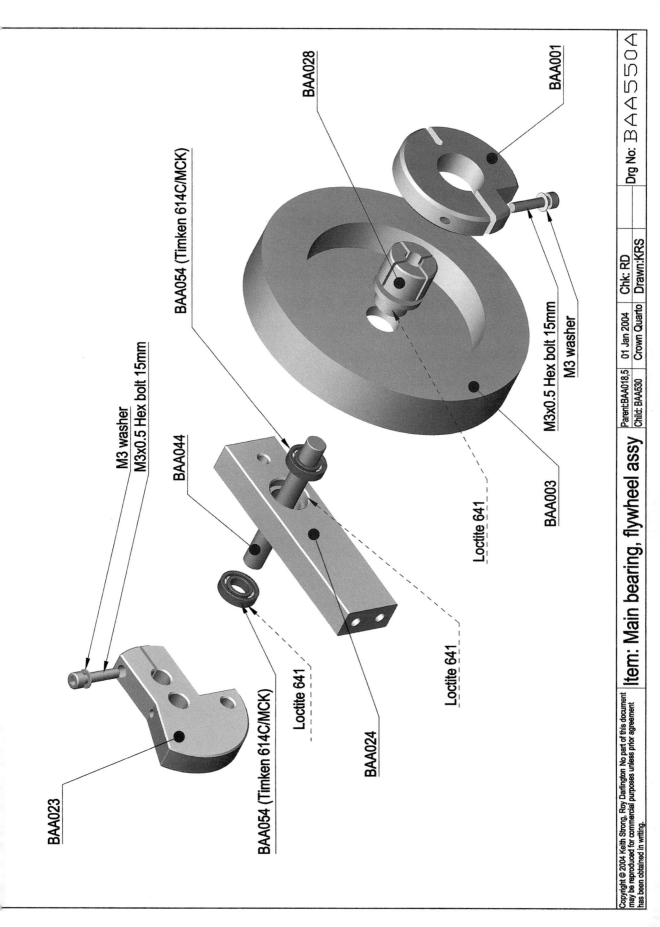

BAA028

BAA001

BAA054 (Timken 614C/MCK)

M3x0.5 Hex bolt 15mm
M3 washer

BAA003

Loctite 641

M3 washer
M3x0.5 Hex bolt 15mm

BAA044

BAA024

Loctite 641

Loctite 641

BAA023

BAA054 (Timken 614C/MCK)

Copyright © 2004 Keith Strong, Roy Darlington No part of this document may be reproduced for commercial purposes unless prior agreement has been obtained in writing.

Item: Main bearing, flywheel assy

| Parent:BAA018,5 | 01 Jan 2004 | Chk: RD | Drg No: BAA550A |
| Child: BAA530 | Crown Quarto | Drawn:KRS | |

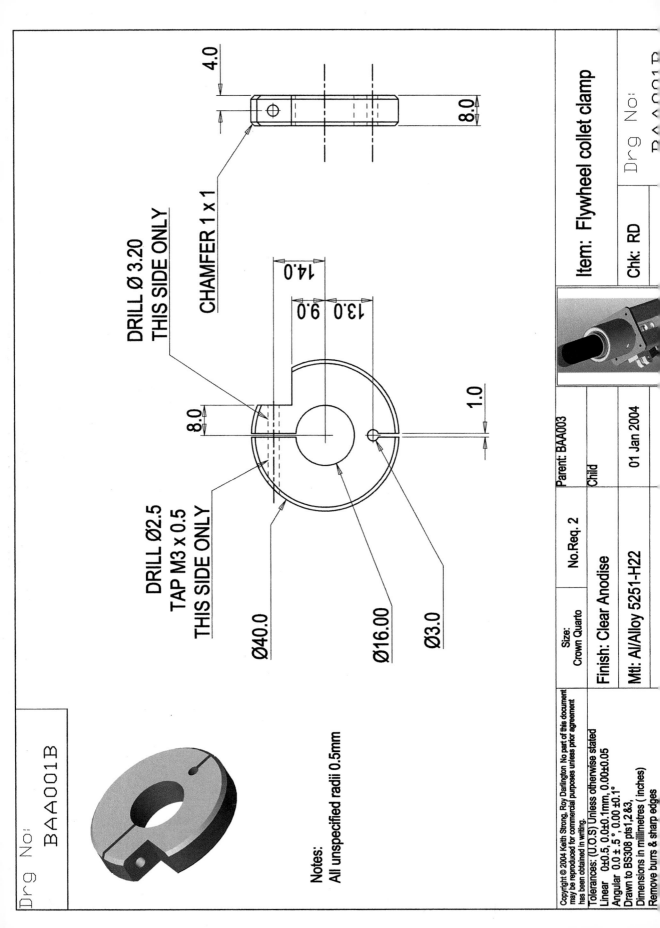

Drg No: BAA001B

4.0

8.0

DRILL Ø 3.20
THIS SIDE ONLY

CHAMFER 1 x 1

14.0

13.0 9.0

8.0

1.0

DRILL Ø2.5
TAP M3 x 0.5
THIS SIDE ONLY

Ø40.0

Ø16.00

Ø3.0

Notes:
All unspecified radii 0.5mm

Item: Flywheel collet clamp

Drg No:

Chk: RD

Parent: BAA003

No.Req. 2

Size:
Crown Quarto

Child

01 Jan 2004

Finish: Clear Anodise

Mtl: Al/Alloy 5251-H22

Copyright © 2004 Keith Strong, Roy Darlington No part of this document may be reproduced for commercial purposes unless prior agreement has been obtained in writing.

Tolerances: (U.O.S) Unless otherwise stated
Linear 0±0.5, 0.0±0.1mm, 0.00±0.05
Angular 0.0 ± 5 ˚, 0.00 ±0.1°
Drawn to BS308 pts1,2 &3,
Dimensions in millimetres (inches)
Remove burrs & sharp edges

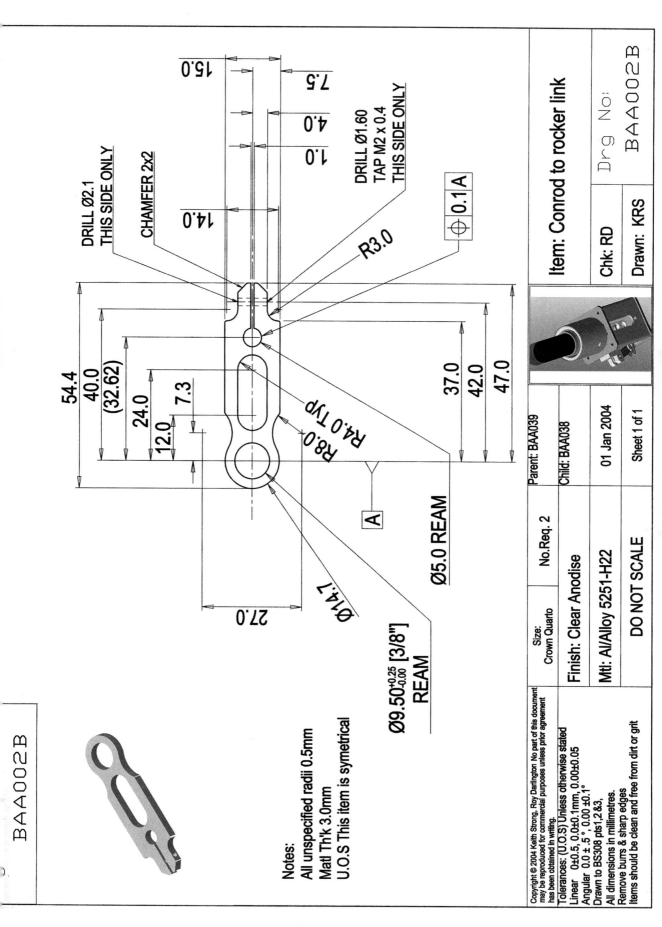

BAA002B

DRILL Ø2.1
THIS SIDE ONLY

CHAMFER 2x2

DRILL Ø1.60
TAP M2 x 0.4
THIS SIDE ONLY

15.0
7.5
4.0
1.0
14.0

R3.0

⌖ 0.1 A

54.4
40.0
(32.62)
24.0
12.0
7.3

R8.0
R4.0 Typ

37.0
42.0
47.0

Ø5.0 REAM

A

27.0

Ø1.7

Ø9.50 +0.25 -0.00 [3/8"]
REAM

Notes:
All unspecified radii 0.5mm
Matl Th'k 3.0mm
U.O.S This item is symetrical

Item: Conrod to rocker link

Drg No:
BAA002B

Chk: RD

Drawn: KRS

Parent: BAA039

Child: BAA038

No.Req. 2

Size:
Crown Quarto

Finish: Clear Anodise

Mtl: Al/Alloy 5251-H22

01 Jan 2004

Sheet 1 of 1

DO NOT SCALE

Copyright © 2004 Keith Strong, Roy Darlington No part of this document
may be reproduced for commercial purposes unless prior agreement
has been obtained in writing.

Tolerances: (U.O.S) Unless otherwise stated
Linear 0±0.5, 0.0±0.1mm, 0.00±0.05
Angular 0.0 ± 5 ° , 0.00 ±0.1°
Drawn to BS308 pts1,2&3,
All dimensions in millimetres.
Remove burrs & sharp edges
Items should be clean and free from dirt or grit

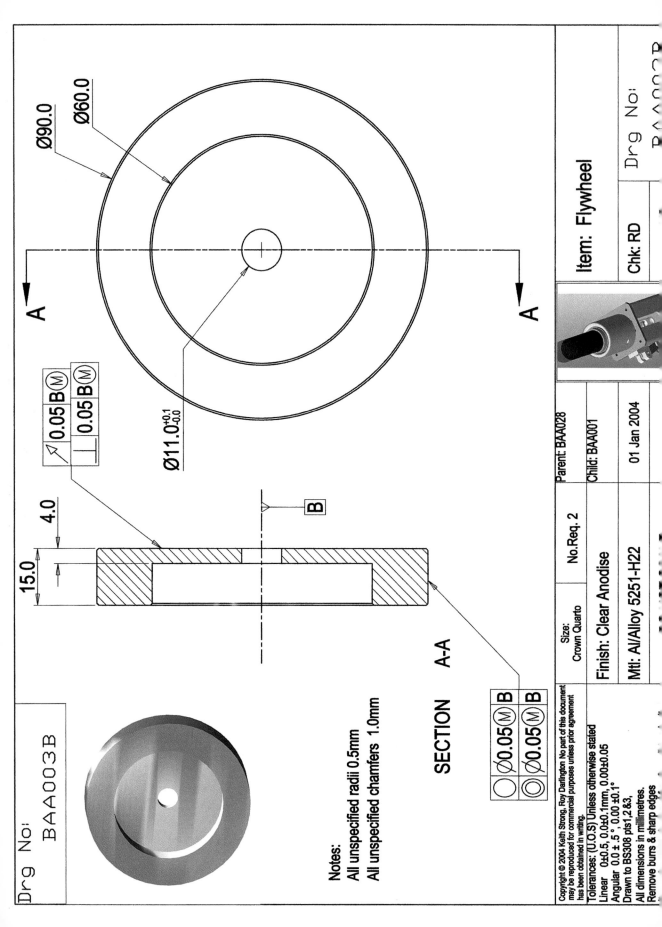

Ø90.0

Ø60.0

Ø11.0$^{+0.1}_{-0.0}$

A

A

| ▽ | 0.05 | B Ⓜ |
| ⊥ | 0.05 | B Ⓜ |

4.0

15.0

B

SECTION A-A

| ◯ | Ø0.05 Ⓜ | B |
| ◎ | Ø0.05 Ⓜ | B |

Drg No:

BAA003B

Notes:
All unspecified radii 0.5mm
All unspecified chamfers 1.0mm

Item: Flywheel

Drg No:
BAA003B

Chk: RD

Parent: BAA028

Child: BAA001

No.Req. 2

01 Jan 2004

Size:
Crown Quarto

Finish: Clear Anodise

Mtl: Al/Alloy 5251-H22

Copyright © 2004 Keith Strong, Roy Darlington No part of this document
may be reproduced for commercial purposes unless prior agreement
has been obtained in writing.

Tolerances: (U.O.S) Unless otherwise stated
Linear   0±0.5, 0.0±0.1mm, 0.00±0.05
Angular 0.0 ± .5˚, 0.00 ±0.1˚
Drawn to BS308 pts1,2 &3,
All dimensions in millimetres.
Remove burrs & sharp edges

BAA004B

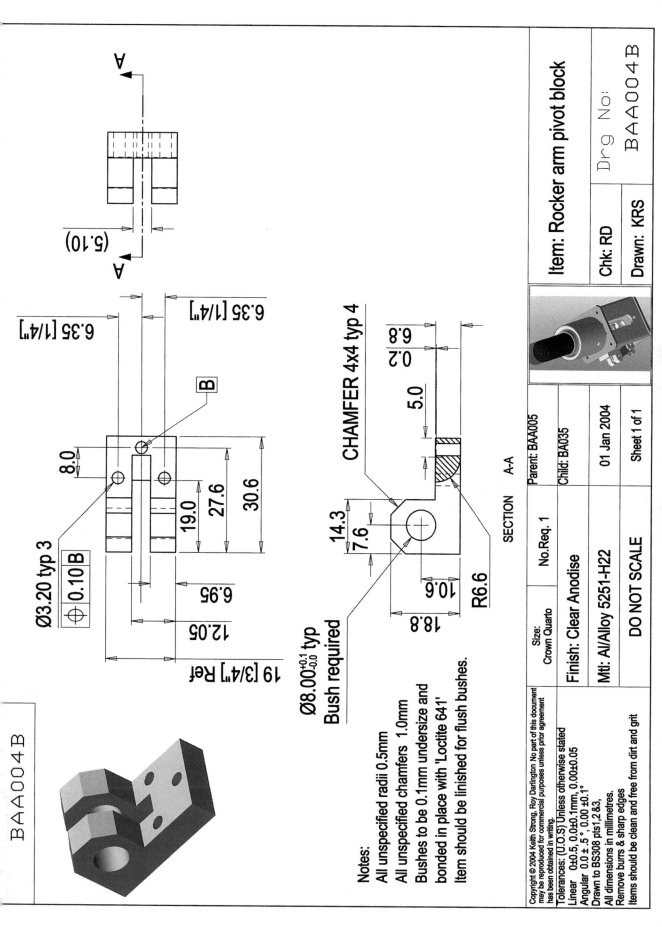

Ø3.20 typ 3

⌖ Ø0.10 B

8.0

19.0
27.6
30.6

6.95

12.05

19 [3/4"] Ref

B

6.35 [1/4"]

6.35 [1/4"]

(5.10)

A

A

Ø8.00 +0.1/-0.0 typ
Bush required

CHAMFER 4x4 typ 4

14.3
7.6

R6.6

10.6

18.8

5.0

0.2
6.8

SECTION   A-A

Notes:
All unspecified radii 0.5mm
All unspecified chamfers 1.0mm
Bushes to be 0.1mm undersize and
bonded in place with 'Loctite 641'
Item should be linished for flush bushes.

Copyright © 2004 Keith Strong, Roy Darlington No part of this document
may be reproduced for commercial purposes unless prior agreement
has been obtained in writing.

Tolerances: (U.O.S) Unless otherwise stated
Linear   0±0.5, 0.0±0.1mm, 0.00±0.05
Angular 0.0 ± .5°, 0.00 ±0.1°
Drawn to BS308 pts1,2 &3,
All dimensions in millimetres.
Remove burrs & sharp edges
Items should be clean and free from dirt and grit

Size:
Crown Quarto

No.Req. 1

Parent: BAA005

Finish: Clear Anodise

Child: BA035

Mtl: Al/Alloy 5251-H22

01 Jan 2004

DO NOT SCALE

Sheet 1 of 1

Item: Rocker arm pivot block

Drg No: BAA004B

Chk: RD

Drawn: KRS

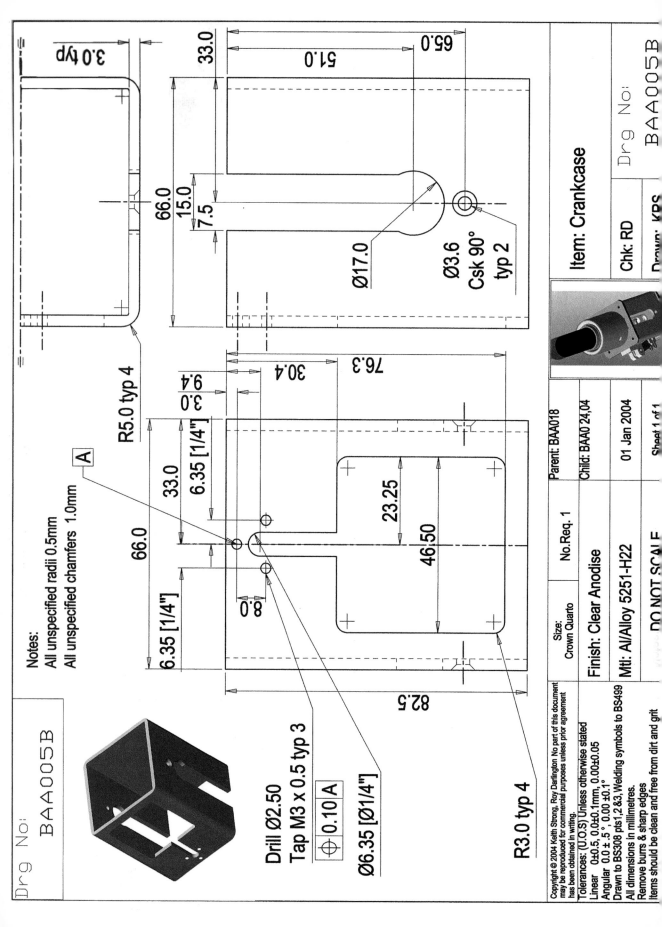

3.0 typ

33.0
51.0
65.0

66.0
15.0
7.5

Ø17.0

Ø3.6
Csk 90°
typ 2

R5.0 typ 4

A

Notes:
All unspecified radii 0.5mm
All unspecified chamfers 1.0mm

76.3
30.4
9.4
3.0

33.0
6.35 [1/4"]

66.0

23.25

46.50

8.0

6.35 [1/4"]

82.5

Drill Ø2.50
Tap M3 x 0.5 typ 3
⊕ 0.10 A
Ø6.35 [Ø1/4"]

R3.0 typ 4

Item: Crankcase

Drg No: BAA005B

Chk: RD

Drawn: KPS

Parent: BAA018

Child: BAA0 24,04

01 Jan 2004

Sheet 1 of 1

Size: Crown Quarto

No.Req. 1

Finish: Clear Anodise

Mtl: Al/Alloy 5251-H22

DO NOT SCALE

Copyright © 2004 Keith Strong, Roy Darlington No part of this document may be reproduced for commercial purposes unless prior agreement has been obtained in writing.
Tolerances: (U.O.S) Unless otherwise stated
Linear 0±0.5, 0.0±0.1mm, 0.00±0.05
Angular 0.0 ± .5°, 0.00 ±0.1°
Drawn to BS308 pts1,2&3, Welding symbols to BS499
All dimensions in millimetres.
Remove burrs & sharp edges
Items should be clean and free from dirt and grit

Drg No: BAA005B

BAA005B

BAA006B

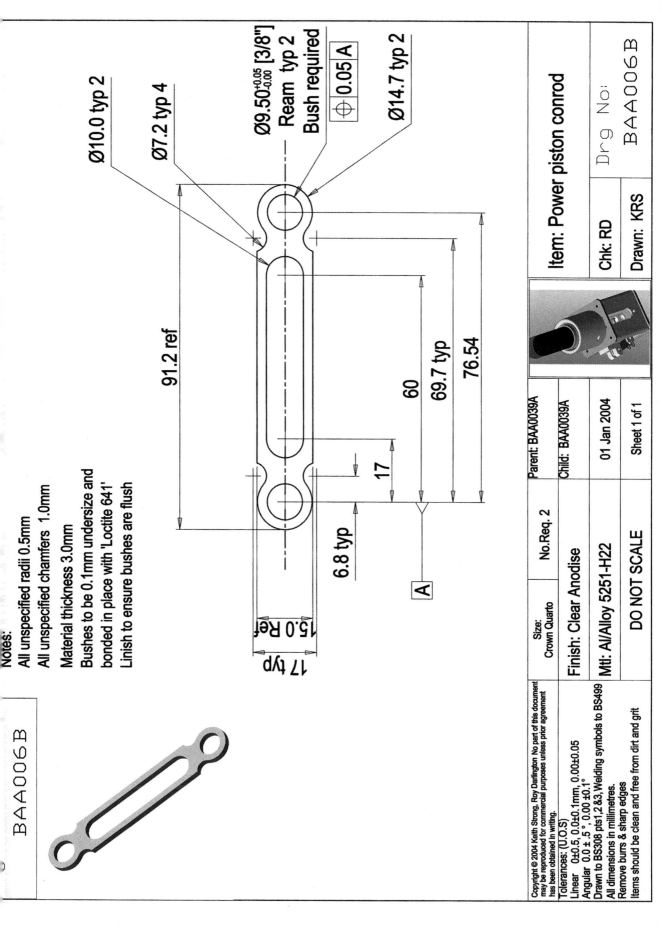

Notes:
All unspecified radii 0.5mm
All unspecified chamfers 1.0mm
Material thickness 3.0mm
Bushes to be 0.1mm undersize and
bonded in place with 'Loctite 641'
Linish to ensure bushes are flush

Ø10.0 typ 2

Ø7.2 typ 4

Ø9.50$^{+0.05}_{-0.00}$ [3/8"]
Ream typ 2
Bush required

⊕ 0.05 A

Ø14.7 typ 2

91.2 ref

76.54

69.7 typ

60

17

6.8 typ

A

15.0 Ref

17 typ

Copyright © 2004 Keith Strong, Roy Darlington No part of this document
may be reproduced for commercial purposes unless prior agreement
has been obtained in writing.

Tolerances: (U.O.S)
Linear   0±0.5, 0.0±0.1mm, 0.00±0.05
Angular 0.0 ± 5°, 0.00 ±0.1°
Drawn to BS308 pts1,2 &3, Welding symbols to BS499
All dimensions in millimetres.
Remove burrs & sharp edges
Items should be clean and free from dirt and grit

Size: Crown Quarto	No.Req. 2	Parent: BAA0039A
		Child: BAA0039A
Finish: Clear Anodise		Item: Power piston conrod
Mtl: Al/Alloy 5251-H22	01 Jan 2004	Chk: RD
DO NOT SCALE	Sheet 1 of 1	Drawn: KRS

Drg No:
BAA006B

This surface to be machined, honed and lapped and matched with the cylinder in accordance with the operation specification in 'Workshop Practice' section of this book.

Chamfer 1x1

1.23

30

3

Ø6.80 +0.018/-0.000 (H7-g6)

◯	⌀0.025 Ⓜ	A
∠	⌀0.025 Ⓜ	A
⟋	⌀0.025 Ⓜ	A

A

Ø33.00 -0.050/-0.075 (H7-e7)

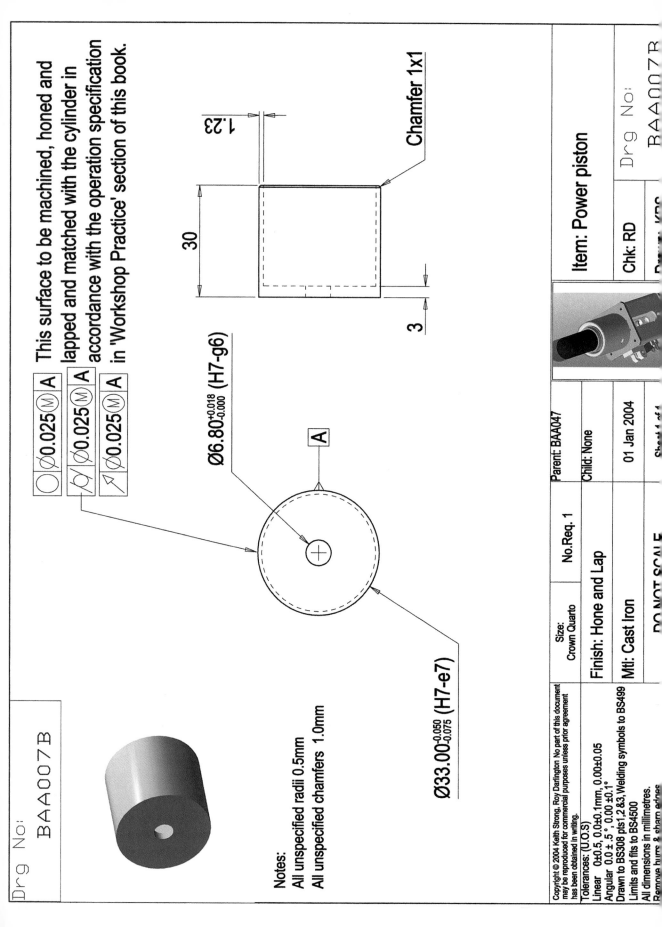

Notes:
All unspecified radii 0.5mm
All unspecified chamfers 1.0mm

Copyright © 2004 Keith Strong, Roy Darlington. No part of this document may be reproduced for commercial purposes unless prior agreement has been obtained in writing.

Tolerances: (U.O.S)
Linear   0±0.5 , 0.0±0.1mm, 0.00±0.05
Angular 0.0 ± .5˚ , 0.00 ±0.1˚
Drawn to BS308 pts1,2 &3, Welding symbols to BS499
Limits and fits to BS4500
All dimensions in millimetres.
Remove burrs & sharp edges

Size: Crown Quarto	No.Req. 1	Parent: BAA047
Finish: Hone and Lap		Child: None
Mtl: Cast Iron		

DO NOT SCALE

Item: Power piston

Chk: RD   Drg No:

Drawn: KDS   BAA007B

01 Jan 2004   Sheet 1 of 1

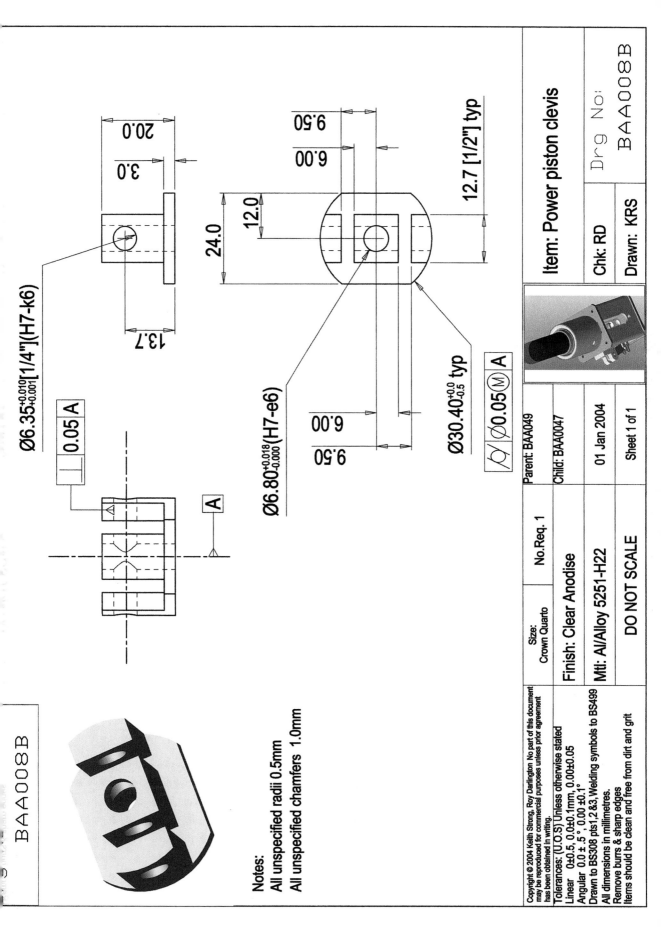

BAA008B

Notes:
All unspecified radii 0.5mm
All unspecified chamfers 1.0mm

20.0
3.0
13.7

$\varnothing 6.35^{+0.010}_{+0.001}[1/4"](H7\text{-}k6)$

⊥ 0.05 A

A

9.50
6.00
12.0
24.0
12.7 [1/2"] typ

$\varnothing 6.80^{+0.018}_{-0.000}(H7\text{-}e6)$

6.00
9.50

$\varnothing 30.40^{+0.0}_{-0.5}$ typ

⌀ $\varnothing 0.05$ Ⓜ A

Item: Power piston clevis

Drg No: BAA008B

Chk: RD

Drawn: KRS

Parent: BAA049

Child: BAA0047

No.Req. 1

Size:
Crown Quarto

Finish: Clear Anodise

Mtl: Al/Alloy 5251-H22

01 Jan 2004

DO NOT SCALE

Sheet 1 of 1

Copyright © 2004 Keith Strong, Roy Darlington No part of this document
may be reproduced for commercial purposes unless prior agreement
has been obtained in writing.

Tolerances: (U.O.S) Unless otherwise stated
Linear   0±0.5, 0.0±0.1mm, 0.00±0.05
Angular 0.0 ± .5°, 0.00 ±0.1°
Drawn to BS308 pts1,2 &3, Welding symbols to BS499
All dimensions in millimetres.
Remove burrs & sharp edges
Items should be clean and free from dirt and grit

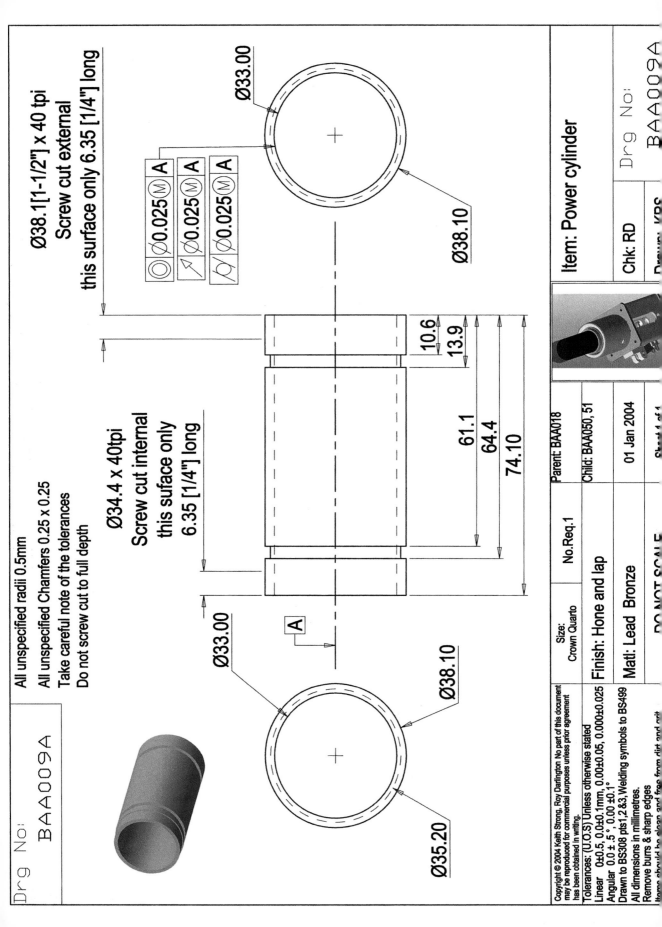

Drg No:
BAA009A

All unspecified radii 0.5mm

All unspecified Chamfers 0.25 x 0.25

Take careful note of the tolerances

Do not screw cut to full depth

Ø38.1[1-1/2"] x 40 tpi
Screw cut external
this surface only 6.35 [1/4"] long

Ø33.00

Ø38.10

⊙	Ø0.025 Ⓜ	A
∠	Ø0.025 Ⓜ	A
⌖	Ø0.025 Ⓜ	A

Ø34.4 x 40tpi
Screw cut internal
this suface only
6.35 [1/4"] long

10.6
13.9
61.1
64.4
74.10

Ø33.00

Ø38.10

Ø35.20

A

A

Copyright © 2004 Keith Strong, Roy Darlington No part of this document may be reproduced for commercial purposes unless prior agreement has been obtained in writing.

Tolerances: (U.O.S) Unless otherwise stated
Linear   0±0.5, 0.0±0.1mm, 0.00±0.05, 0.000±0.025
Angular 0.0 ± .5°, 0.00 ±0.1°
Drawn to BS308 pts1,2 &3, Welding symbols to BS499
All dimensions in millimetres.
Remove burrs & sharp edges
Item should be clean and free from dirt and grit

Size:
Crown Quarto

No.Req.1

Parent: BAA018

Child: BAA050, 51

Finish: Hone and lap

Matl: Lead Bronze

DO NOT SCALE

01 Jan 2004

Item: Power cylinder

Chk: RD

Drg No:
BAA009A

Drawn: KPS

Sheet 1 of 1

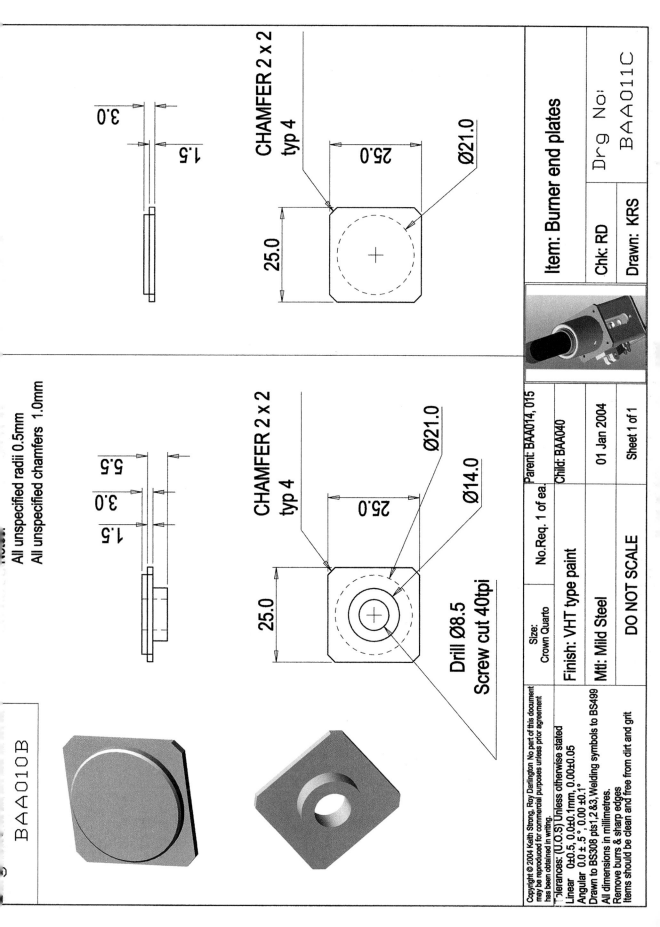

BAA010B

All unspecified radii 0.5mm
All unspecified chamfers 1.0mm

3.0

1.5

5.5
3.0
1.5

CHAMFER 2 x 2
typ 4

Ø21.0

25.0

25.0

CHAMFER 2 x 2
typ 4

Ø21.0
Ø14.0

25.0

25.0

Drill Ø8.5
Screw cut 40tpi

Item: Burner end plates

Drg No: BAA011C

Chk: RD

Drawn: KRS

Copyright © 2004 Keith Strong, Roy Darlington No part of this document may be reproduced for commercial purposes unless prior agreement has been obtained in writing.

Tolerances: (U.O.S) Unless otherwise stated
Linear   0±0.5, 0.0±0.1mm, 0.00±0.05
Angular 0.0 ± .5°, 0.00 ±0.1°
Drawn to BS308 pts1,2 &3, Welding symbols to BS499
All dimensions in millimetres.
Remove burrs & sharp edges
Items should be clean and free from dirt and grit

Size: Crown Quarto

No.Req. 1 of ea.

Parent: BAA014, 015
Child: BAA040

Finish: VHT type paint

Mtl: Mild Steel

01 Jan 2004

DO NOT SCALE

Sheet 1 of 1

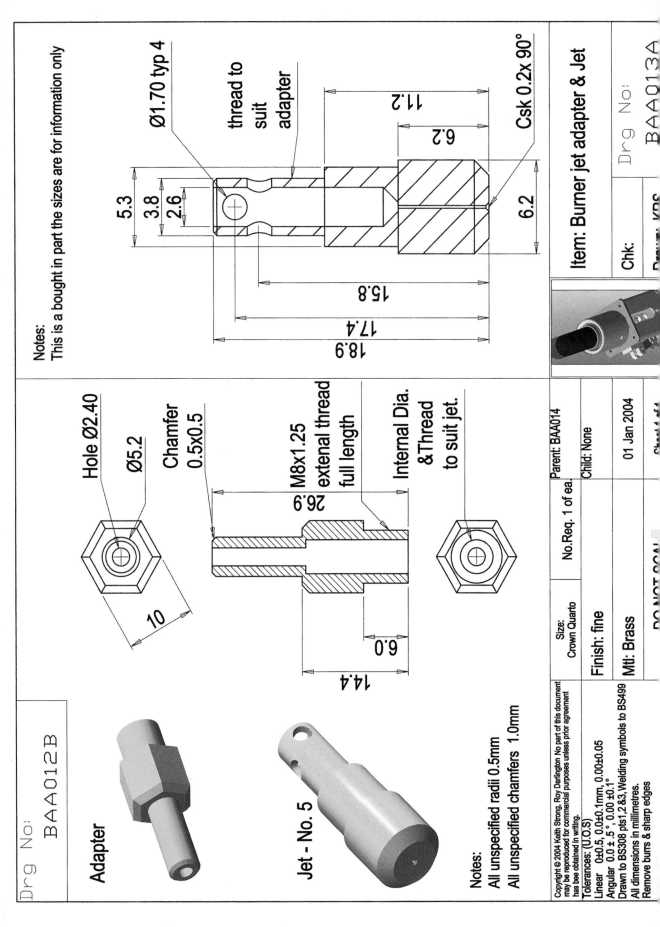

Notes:
This is a bought in part the sizes are for information only

Ø1.70 typ 4

thread to suit adapter

Csk 0.2x 90°

5.3
3.8
2.6

11.2
6.2
6.2

15.8

17.4

18.9

Drg No:
BAA012B

Hole Ø2.40

Ø5.2

Chamfer 0.5x0.5

M8x1.25 extenal thread full length

Internal Dia. &Thread to suit jet.

26.9

10

6.0

14.4

Adapter

Jet - No. 5

Notes:
All unspecified radii 0.5mm
All unspecified chamfers 1.0mm

Item: Burner jet adapter & Jet

Drg No:
BAA013A

Chk:

Drawn: KDS

Copyright © 2004 Keith Strong, Roy Darlington No part of this document may be reproduced for commercial purposes unless prior agreement has bee obtained in writing.

Tolerances: (U.O.S)
Linear 0±0.5, 0.0±0.1mm, 0.00±0.05
Angular 0.0 ± .5°, 0.00 ±0.1°
Drawn to BS308 pts1,2 &3, Welding symbols to BS499
All dimensions in millimetres.
Remove burrs & sharp edges

Size:
Crown Quarto

No.Req. 1 of ea.

Parent: BAA014

Child: None

Finish: fine

Mtl: Brass

01 Jan 2004

Sheet 1 of 1

DO NOT SCALE

BAA014A

Notes:
All unspecified radii 0.5mm
All unspecified chamfers 1.0mm

Ø 8.40

6.35

76.0

Cross Drill 90°
Ø5.3 typ 4

16.7

DRILL AND TAP
M8x1.25 INTERNAL
7.00 DEEP

Ø9.1

Ø7.0

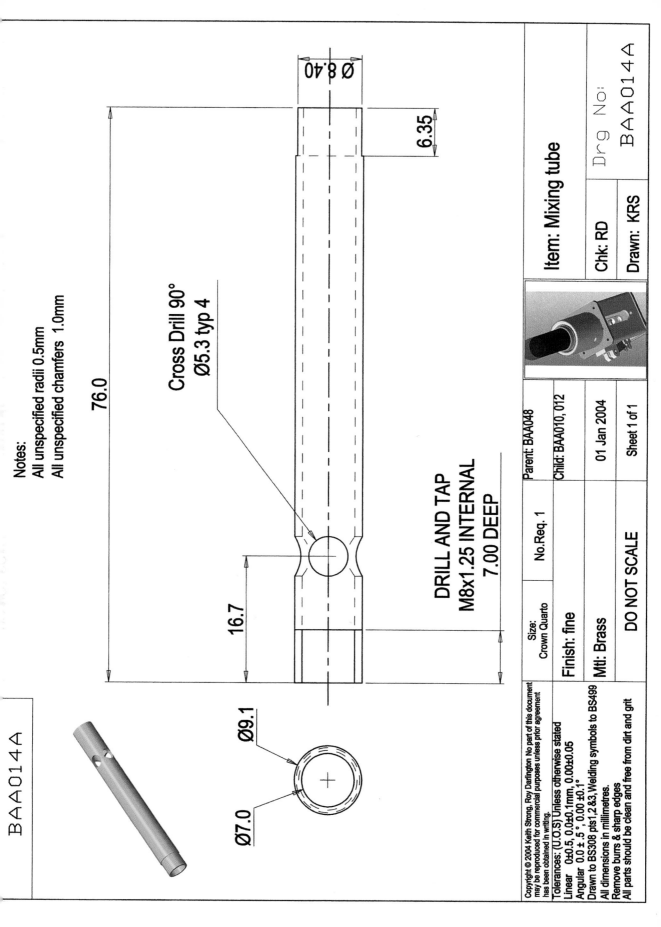

Item: Mixing tube

Copyright © 2004 Keith Strong, Roy Darlington No part of this document may be reproduced for commercial purposes unless prior agreement has been obtained in writing.

	Size: Crown Quarto	Parent: BAA048	
		No.Req. 1	Child: BAA010, 012

Tolerances: (U.O.S) Unless otherwise stated
Linear   0±0.5, 0.0±0.1mm, 0.00±0.05
Angular 0.0 ± .5°, 0.00 ±0.1°
Drawn to BS308 pts1,2 &3, Welding symbols to BS499
All dimensions in millimetres.
Remove burrs & sharp edges
All parts should be clean and free from dirt and grit

Finish: fine

Mtl: Brass

DO NOT SCALE

01 Jan 2004

Sheet 1 of 1

Chk: RD

Drg No:

BAA014A

Drawn: KRS

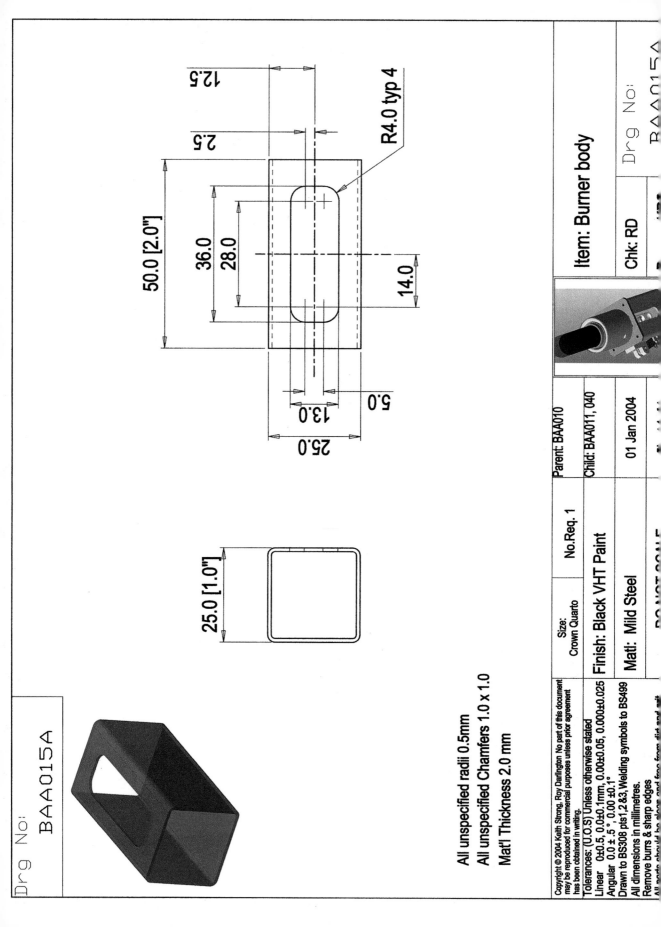

12.5

2.5

R4.0 typ 4

50.0 [2.0"]

36.0

28.0

14.0

13.0

5.0

25.0

25.0 [1.0"]

All unspecified radii 0.5mm
All unspecified Chamfers 1.0 x 1.0
Mat'l Thickness 2.0 mm

Item: Burner body

Chk: RD

Parent: BAA010

Child: BAA011, 040

Size:
Crown Quarto

No.Req. 1

Finish: Black VHT Paint

Mat'l: Mild Steel

01 Jan 2004

Copyright © 2004 Keith Strong, Roy Darlington. No part of this document
may be reproduced for commercial purposes unless prior agreement
has been obtained in writing.

Tolerances: (U.O.S) Unless otherwise stated
Linear 0±0.5, 0.0±0.1mm, 0.00±0.05, 0.000±0.025
Angular 0.0 ± .5°, 0.00 ±0.1°
Drawn to BS308 pts1,2 &3, Welding symbols to BS499
All dimensions in millimetres.
Remove burrs & sharp edges
All parts should be clean and free from dirt and oil

BAA016A

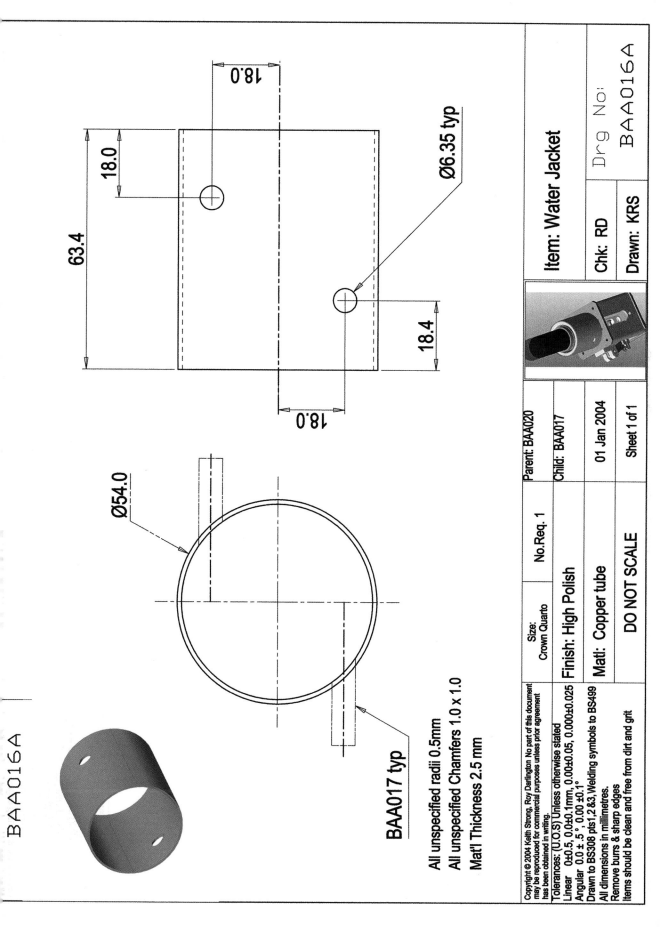

18.0

18.0

63.4

Ø6.35 typ

18.4

18.0

Ø54.0

BAA017 typ

All unspecified radii 0.5mm
All unspecified Chamfers 1.0 x 1.0
Mat'l Thickness 2.5 mm

Copyright © 2004 Keith Strong, Roy Darlington No part of this document may be reproduced for commercial purposes unless prior agreement has been obtained in writing.

Tolerances: (U.O.S) Unless otherwise stated
Linear 0±0.5, 0.0±0.1mm, 0.00±0.05, 0.000±0.025
Angular 0.0 ± .5˚ , 0.00 ±0.1˚
Drawn to BS308 pts1,2&3, Welding symbols to BS499
All dimensions in millimetres.
Remove burrs & sharp edges
Items should be clean and free from dirt and grit

Size: Crown Quarto	No.Req. 1	Parent: BAA020
Finish: High Polish		Child: BAA017
Matl: Copper tube		01 Jan 2004
DO NOT SCALE		Sheet 1 of 1

Item: Water Jacket

Chk: RD

Drg No: BAA016A

Drawn: KRS

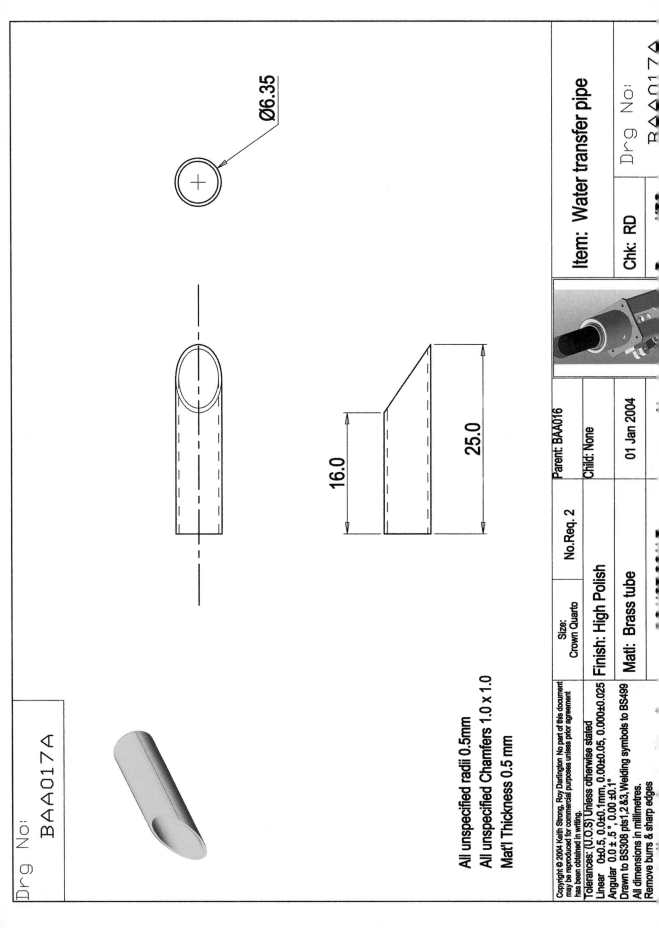

Drg No:
BAA017A

Ø6.35

16.0

25.0

All unspecified radii 0.5mm
All unspecified Chamfers 1.0 x 1.0
Mat'l Thickness 0.5 mm

Copyright © 2004 Keith Strong, Roy Darlington No part of this document may be reproduced for commercial purposes unless prior agreement has been obtained in writing.

Tolerances: (U.O.S) Unless otherwise stated
Linear 0±0.5, 0.0±0.1mm, 0.00±0.05, 0.000±0.025
Angular 0.0 ± 5° , 0.00 ±0.1°
Drawn to BS308 pts1,2 &3, Welding symbols to BS499
All dimensions in millimetres.
Remove burrs & sharp edges

Size:
Crown Quarto

No.Req. 2

Parent: BAA016

Child: None

Finish: High Polish

Matl: Brass tube

01 Jan 2004

Item: Water transfer pipe

Chk: RD

Drg No:
BAA017A

Drg No:
BAA017A

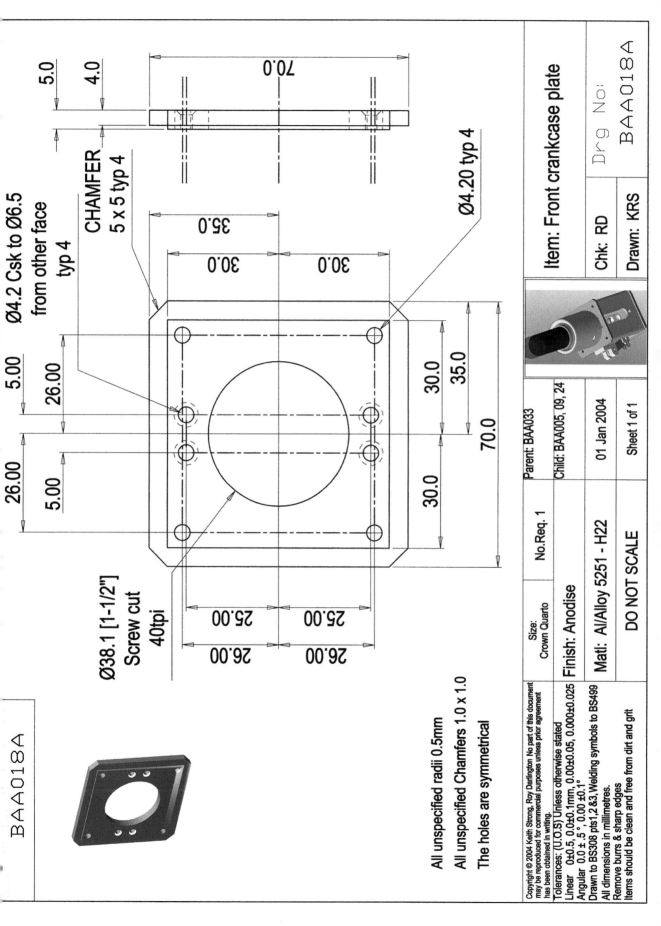

BAA018A

Ø4.2 Csk to Ø6.5
from other face
typ 4

CHAMFER
5 x 5 typ 4

Ø4.20 typ 4

5.0
4.0

70.0

35.0
30.0
30.0

5.00
26.00
26.00
5.00

30.0
35.0
30.0
70.0

Ø38.1 [1-1/2"]
Screw cut
40tpi

25.00
25.00
26.00
26.00

All unspecified radii 0.5mm
All unspecified Chamfers 1.0 x 1.0
The holes are symmetrical

Copyright © 2004 Keith Strong, Roy Darlington No part of this document may be reproduced for commercial purposes unless prior agreement has been obtained in writing.

Tolerances: (U.O.S) Unless otherwise stated
Linear 0±0.5, 0.0±0.1mm, 0.00±0.05, 0.000±0.025
Angular 0.0 ± .5˚ , 0.00 ±0.1˚
Drawn to BS308 pts1,2 &3, Welding symbols to BS499
All dimensions in millimetres.
Remove burrs & sharp edges
Items should be clean and free from dirt and grit

Size:
Crown Quarto

Finish: Anodise

Matl: Al/Alloy 5251 - H22

DO NOT SCALE

No.Req. 1

Parent: BAA033
Child: BAA005, 09, 24

01 Jan 2004

Sheet 1 of 1

Item: Front crankcase plate

Drg No:
BAA018A

Chk: RD

Drawn: KRS

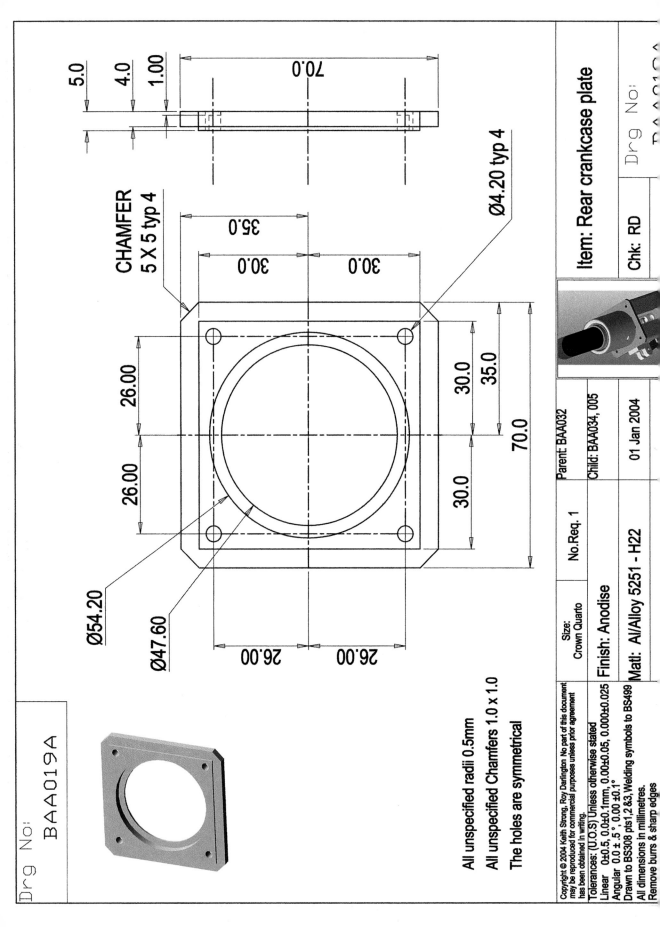

5.0
4.0
1.00

70.0

CHAMFER
5 X 5 typ 4

Ø4.20 typ 4

35.0
30.0
30.0

26.00
26.00

30.0
35.0
70.0
30.0

Ø54.20
Ø47.60

26.00
26.00

All unspecified radii 0.5mm
All unspecified Chamfers 1.0 x 1.0
The holes are symmetrical

Copyright © 2004 Keith Strong, Roy Darlington No part of this document may be reproduced for commercial purposes unless prior agreement has been obtained in writing.

Tolerances: (U.O.S) Unless otherwise stated
Linear   0±0.5, 0.0±0.1mm, 0.00±0.05, 0.000±0.025
Angular 0.0 ± .5°, 0.00 ±0.1°
Drawn to BS308 pts1,2 &3, Welding symbols to BS499
All dimensions in millimetres.
Remove burrs & sharp edges

Size:
Crown Quarto

No.Req. 1

Parent: BAA032

Child: BAA034, 005

Finish: Anodise

Matl: Al/Alloy 5251 - H22

01 Jan 2004

Item: Rear crankcase plate

Drg No:

Chk: RD

BAA020A

11

Ø51.50

Ø41.7

R2

All unspecified radii 0.5mm

All unspecified Chamfers 1.0 x 1.0

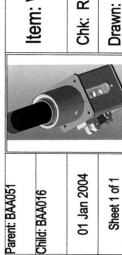

Item: Water jacket end seals

Drg No: BAA020A

| | Chk: RD |
| Drawn: KRS | |

Copyright © 2004 Keith Strong, Roy Darlington No part of this document may be reproduced for commercial purposes unless prior agreement has been obtained in writing.

Parent: BAA051

No.Req. 2

Size: Crown Quarto

Tolerances: (U.O.S) Unless otherwise stated
Linear 0±0.5, 0.0±0.1mm, 0.00±0.05, 0.000±0.025
Angular 0.0 ± .5°, 0.00 ±0.1°
Drawn to BS308 pts1,2 &3, Welding symbols to BS499
All dimensions in millimetres.
Remove burrs & sharp edges
Items should be clean and free from dirt or grit

Finish: Anodise

Child: BAA016

Matl: Al/Alloy 5251 - H22

01 Jan 2004

DO NOT SCALE

Sheet 1 of 1

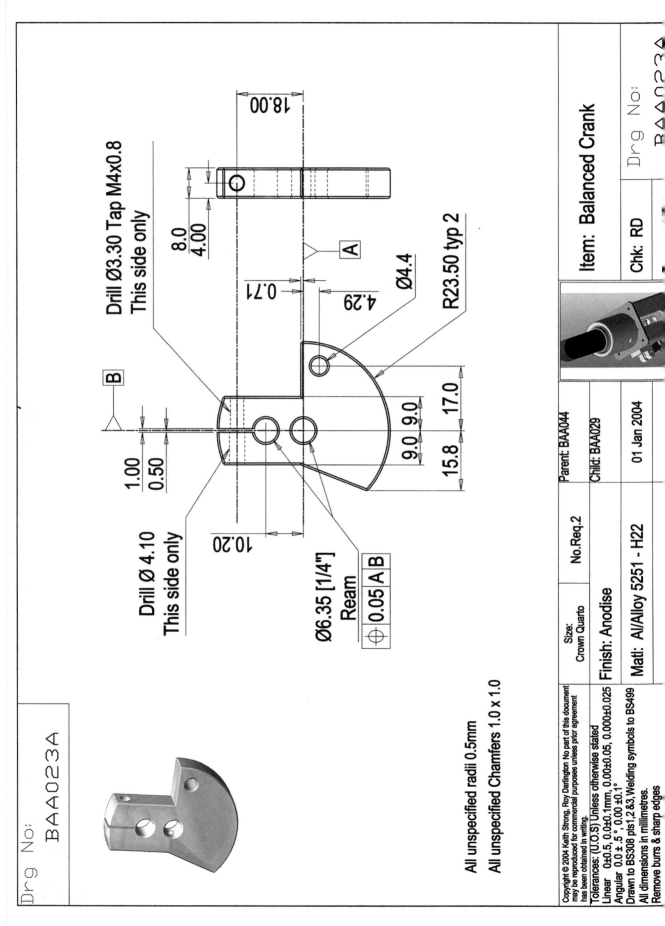

Drill Ø3.30 Tap M4x0.8
This side only

18.00

8.0
4.00

0.71

4.29

Ø4.4

R23.50 typ 2

A

B

17.0

9.0 9.0

15.8

1.00
0.50

10.20

Drill Ø 4.10
This side only

Ø6.35 [1/4"]
Ream
⊕ | 0.05 | A | B

All unspecified radii 0.5mm
All unspecified Chamfers 1.0 x 1.0

Copyright © 2004 Keith Strong, Roy Darlington No part of this document may be reproduced for commercial purposes unless prior agreement has been obtained in writing.

Parent: BAA044

Child: BAA029

No.Req.2

Size:
Crown Quarto

Item: Balanced Crank

Finish: Anodise

Matl: Al/Alloy 5251 - H22

01 Jan 2004

Tolerances: (U.O.S) Unless otherwise stated
Linear   0±0.5, 0.0±0.1mm, 0.00±0.05, 0.000±0.025
Angular 0.0 ± .5°, 0.00 ±0.1°
Drawn to BS308 pts1,2 &3, Welding symbols to BS499
All dimensions in millimetres.
Remove burrs & sharp edges

Drg No:
BAA023A

Chk: RD

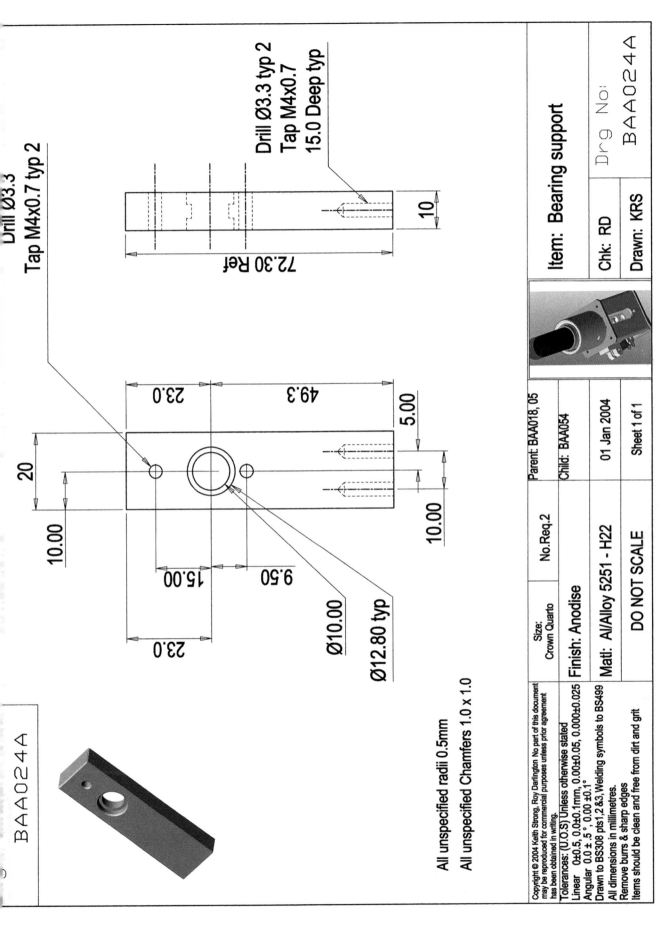

Drill Ø3.3
Tap M4x0.7 typ 2

Drill Ø3.3 typ 2
Tap M4x0.7
15.0 Deep typ

72.30 Ref

10

BAA024A

20

10.00

23.0

15.00

9.50

23.0

49.3

5.00

10.00

Ø10.00

Ø12.80 typ

All unspecified radii 0.5mm

All unspecified Chamfers 1.0 x 1.0

Item: **Bearing support**

Drg No:
BAA024A

Chk: RD

Drawn: KRS

Parent: BAA018, 05

Child: BAA054

No.Req.2

Size:
Crown Quarto

Finish: Anodise

Matl: Al/Alloy 5251 - H22

01 Jan 2004

Sheet 1 of 1

DO NOT SCALE

Copyright © 2004 Keith Strong, Roy Darlingdon No part of this document may be reproduced for commercial purposes unless prior agreement has been obtained in writing.

Tolerances: (U.O.S) Unless otherwise stated
Linear 0±0.5, 0.0±0.1mm, 0.00±0.05, 0.000±0.025
Angular 0.0 ± .5˚ , 0.00 ±0.1˚
Drawn to BS308 pts1,2 &3, Welding symbols to BS499
All dimensions in millimetres.
Remove burrs & sharp edges
Items should be clean and free from dirt and grit

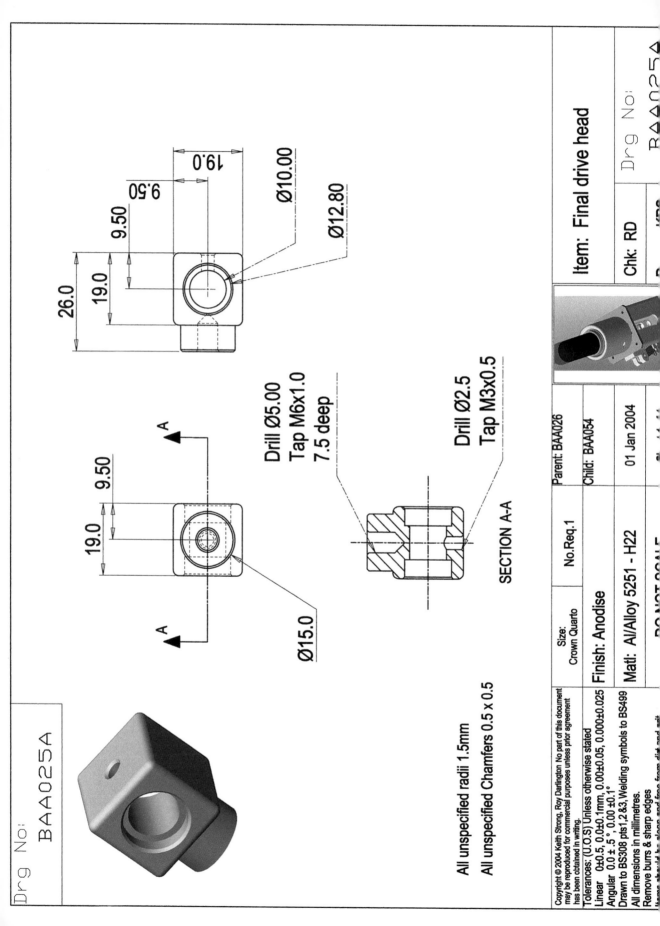

Drg No:
BAA025A

Drg No:
BAA025A

19.0
9.50
9.50
19.0
26.0

Ø10.00
Ø12.80

A
A
9.50
19.0
Ø15.0

Drill Ø5.00
Tap M6x1.0
7.5 deep

Drill Ø2.5
Tap M3x0.5

SECTION A-A

All unspecified radii 1.5mm
All unspecified Chamfers 0.5 x 0.5

Item: Final drive head

Drg No:
BAA025A

Chk: RD

Parent: BAA026

No.Req.1

Child: BAA054

01 Jan 2004

Size:
Crown Quarto

Finish: Anodise

Matl: Al/Alloy 5251 - H22

Copyright © 2004 Keith Strong, Roy Darlington No part of this document
may be reproduced for commercial purposes unless prior agreement
has been obtained in writing.

Tolerances: (U.O.S) Unless otherwise stated
Linear   0±0.5, 0.0±0.1mm, 0.00±0.05, 0.000±0.025
Angular 0.0 ± .5 °, 0.00 ±0.1°
Drawn to BS308 pts1,2 &3, Welding symbols to BS499
All dimensions in millimetres.
Remove burrs & sharp edges

BAA026A

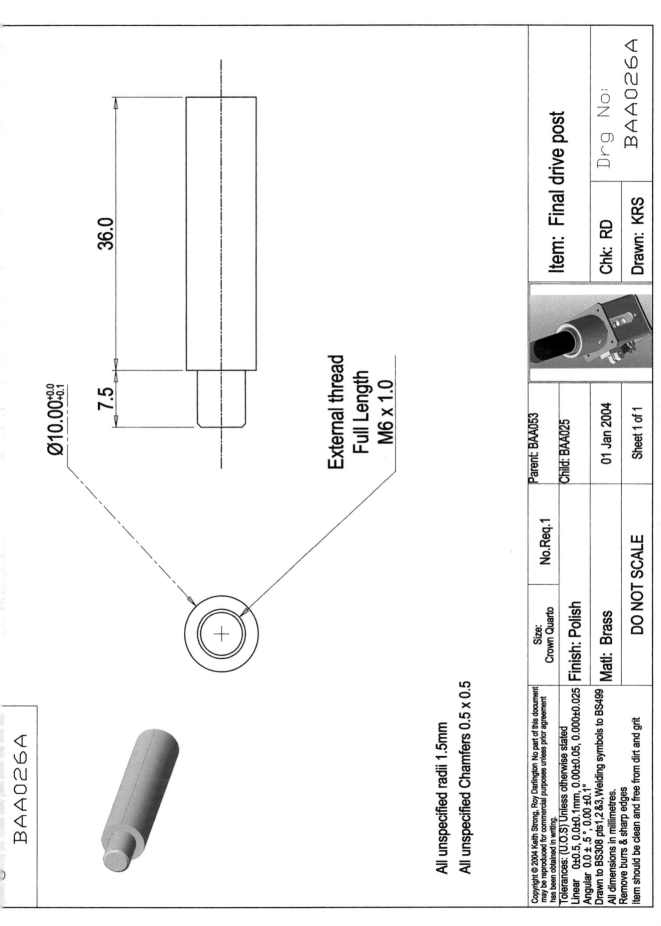

36.0

7.5

$\varnothing 10.00^{+0.0}_{+0.1}$

External thread
Full Length
M6 x 1.0

All unspecified radii 1.5mm

All unspecified Chamfers 0.5 x 0.5

Copyright © 2004 Keith Strong, Roy Darlington No part of this document may be reproduced for commercial purposes unless prior agreement has been obtained in writing.

Tolerances: (U.O.S) Unless otherwise stated
Linear   0±0.5, 0.0±0.1mm, 0.00±0.05, 0.000±0.025
Angular 0.0 ± 5°, 0.00 ±0.1°
Drawn to BS308 pts1,2 &3, Welding symbols to BS499
All dimensions in millimetres.
Remove burrs & sharp edges
Item should be clean and free from dirt and grit

Size: Crown Quarto	No.Req.1	Parent: BAA053
Finish: Polish		Child: BAA025
Matl: Brass		01 Jan 2004
DO NOT SCALE		Sheet 1 of 1

Item: Final drive post

Drg No: BAA026A

Chk: RD

Drawn: KRS

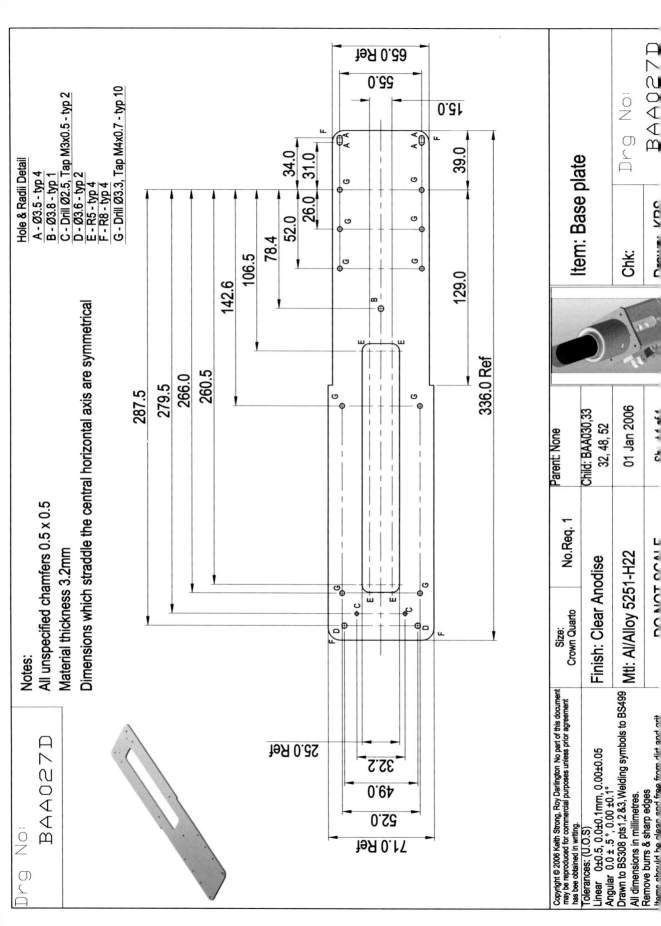

Dng No:
BAA027D

Notes:

All unspecified chamfers 0.5 x 0.5

Material thickness 3.2mm

Dimensions which straddle the central horizontal axis are symmetrical

Hole & Radii Detail
A - Ø3.5 - typ 4
B - Ø3.8 - typ 1
C - Drill Ø2.5, Tap M3x0.5 - typ 2
D - Ø3.6 - typ 2
E - R5 - typ 4
F - R8 - typ 4
G - Drill Ø3.3, Tap M4x0.7 - typ 10

65.0 Ref
55.0
15.0
34.0
31.0
26.0
52.0
78.4
106.5
142.6
287.5
279.5
266.0
260.5
39.0
129.0
336.0 Ref
25.0 Ref
32.2
49.0
52.0
71.0 Ref

Item: **Base plate**

Chk:

Dng No:
BAA027D

Parent: None

Child: BAA030,33
32, 48, 52

01 Jan 2006

Size:
Crown Quarto

No.Req. 1

Finish: Clear Anodise

Mtl: Al/Alloy 5251-H22

Copyright © 2006 Keith Strong, Roy Darlington No part of this document may be reproduced for commercial purposes unless prior agreement has bee obtained in writing.

Tolerances: (U.O.S)
Linear   0±0.5, 0.0±0.1mm, 0.00±0.05
Angular 0.0 ± .5°, 0.00 ±0.1°
Drawn to BS308 pts1,2 &3, Welding symbols to BS499
All dimensions in millimetres.
Remove burrs & sharp edges
Items should be clean and free from dirt and grit

BAA028A

All unspecified radii 1.5mm

All unspecified Chamfers 0.5 x 0.5

Ø11.00

Ø6.35 [1/4"]
Ream

A

A

SECTION   A-A

19.50

16.50

9.50

7.50

5.50

1.00 typ

Ø16.0

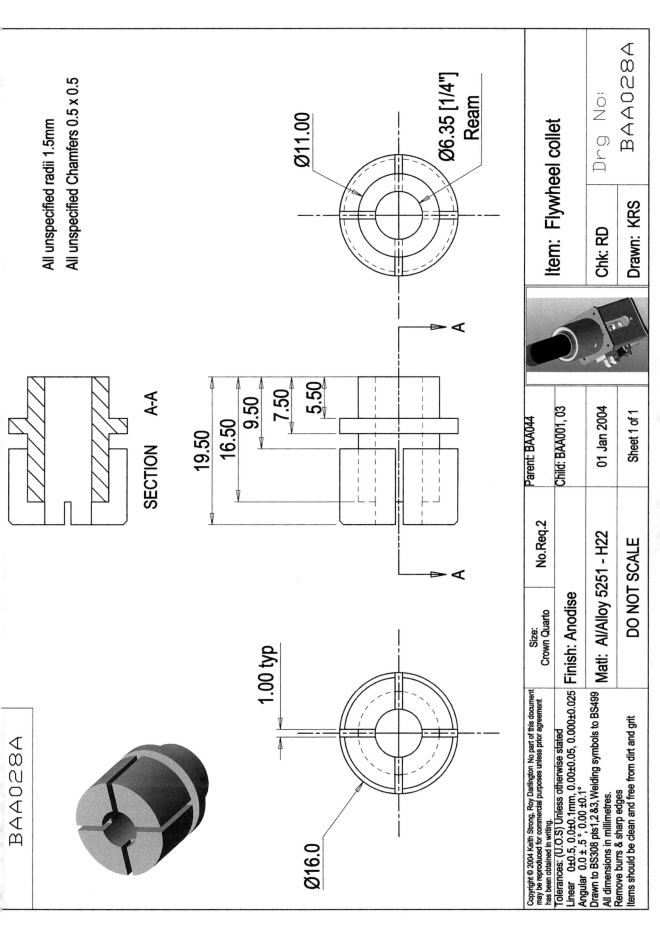

Copyright © 2004 Keith Strong, Roy Darlington No part of this document may be reproduced for commercial purposes unless prior agreement has been obtained in writing.

Tolerances: (U.O.S) Unless otherwise stated
Linear 0±0.5, 0.0±0.1mm, 0.00±0.05, 0.000±0.025
Angular 0.0 ±.5˚, 0.00 ±0.1˚
Drawn to BS308 pts1,2 &3, Welding symbols to BS499
All dimensions in millimetres.
Remove burrs & sharp edges
Items should be clean and free from dirt and grit

Size:
Crown Quarto

No.Req.2

Parent: BAA044

Child: BAA001, 03

Finish: Anodise

Matl: Al/Alloy 5251 - H22

DO NOT SCALE

01 Jan 2004

Sheet 1 of 1

Item: Flywheel collet

Drg No:

BAA028A

Chk: RD

Drawn: KRS

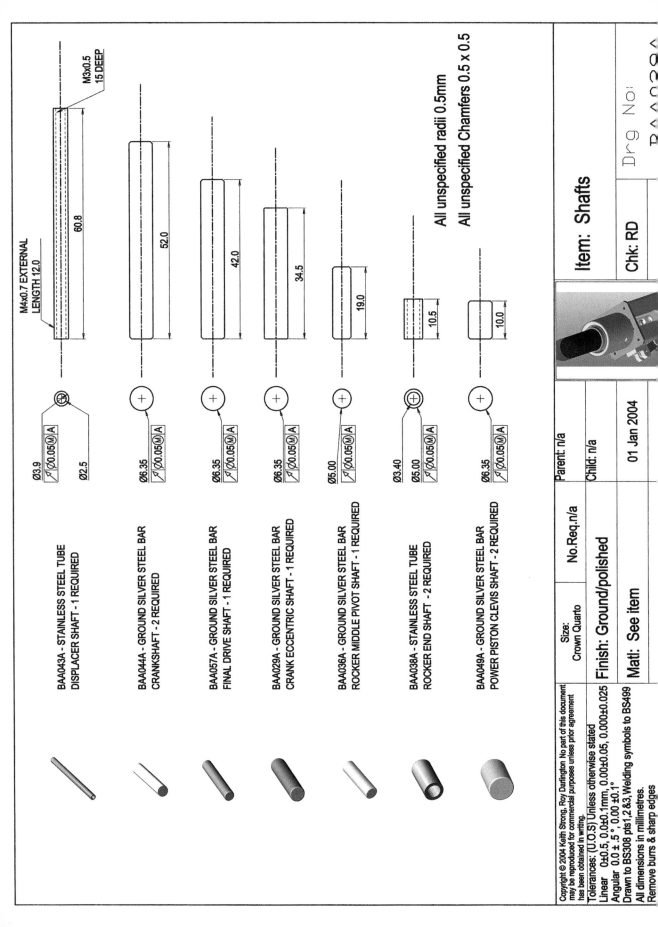

M4x0.7 EXTERNAL
LENGTH 12.0

M3x0.5
15 DEEP

60.8

Ø3.9
⌀0.05 Ⓜ A
Ø2.5

BAA043A - STAINLESS STEEL TUBE
DISPLACER SHAFT - 1 REQUIRED

52.0

Ø6.35
⌀0.05 Ⓜ A

BAA044A - GROUND SILVER STEEL BAR
CRANKSHAFT - 2 REQUIRED

42.0

Ø6.35
⌀0.05 Ⓜ A

BAA057A - GROUND SILVER STEEL BAR
FINAL DRIVE SHAFT - 1 REQUIRED

34.5

Ø6.35
⌀0.05 Ⓜ A

BAA029A - GROUND SILVER STEEL BAR
CRANK ECCENTRIC SHAFT - 1 REQUIRED

19.0

Ø5.00
⌀0.05 Ⓜ A

BAA036A - GROUND SILVER STEEL BAR
ROCKER MIDDLE PIVOT SHAFT - 1 REQUIRED

10.5

Ø3.40
Ø5.00
⌀0.05 Ⓜ A

BAA038A - STAINLESS STEEL TUBE
ROCKER END SHAFT - 2 REQUIRED

10.0

Ø6.35
⌀0.05 Ⓜ A

BAA049A - GROUND SILVER STEEL BAR
POWER PISTON CLEVIS SHAFT - 2 REQUIRED

All unspecified radii 0.5mm

All unspecified Chamfers 0.5 x 0.5

Item: Shafts

Drg No: BAA029A

Chk: RD

Parent: n/a

Child: n/a

No.Req.n/a

Size:
Crown Quarto

Finish: Ground/polished

Matl: See item

01 Jan 2004

Copyright © 2004 Keith Strong, Roy Darlington No part of this document may be reproduced for commercial purposes unless prior agreement has been obtained in writing.

Tolerances: (U.O.S) Unless otherwise stated
Linear 0±0.5, 0.0±0.1mm, 0.00±0.05, 0.000±0.025
Angular 0.0 ± .5˚ , 0.00 ±0.1˚
Drawn to BS308 pts1,2 &3, Welding symbols to BS499
All dimensions in millimetres.
Remove burrs & sharp edges

BAAO30B

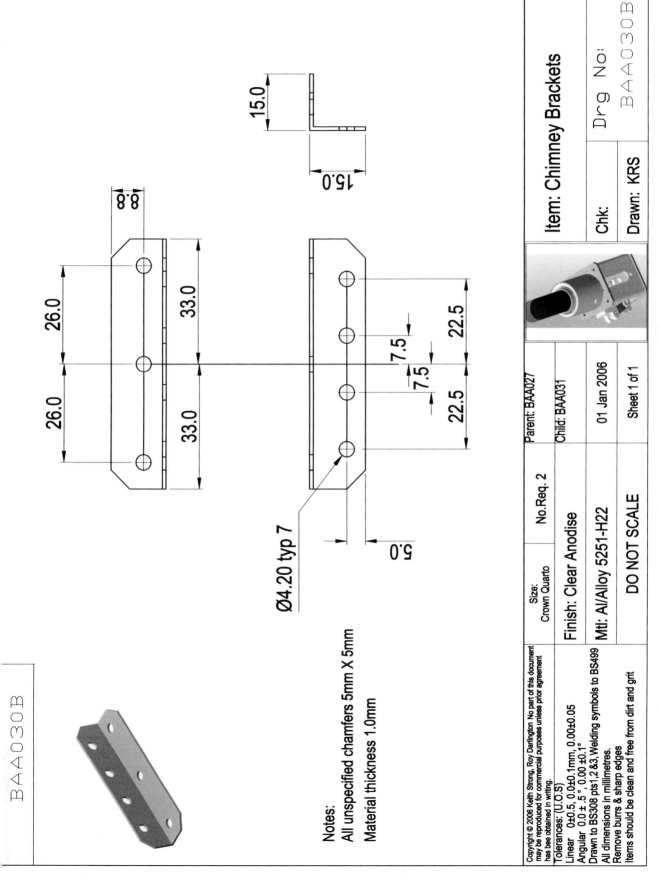

15.0

15.0

8.8

26.0

33.0

26.0

33.0

7.5

7.5

22.5

22.5

5.0

Ø4.20 typ 7

Notes:
All unspecified chamfers 5mm X 5mm

Material thickness 1.0mm

Copyright © 2006 Keith Strong, Roy Darlington No part of this document
may be reproduced for commercial purposes unless prior agreement
has bee obtained in writing.

Tolerances: (U.O.S)
Linear 0±0.5, 0.0±0.1mm, 0.00±0.05
Angular 0.0 ± .5°, 0.00 ±0.1°
Drawn to BS308 pts1,2 &3,Welding symbols to BS499
All dimensions in millimetres.
Remove burrs & sharp edges
Items should be clean and free from dirt and grit

Parent: BAA027

Child: BAA031

Size:
Crown Quarto

No.Req. 2

Finish: Clear Anodise

Mtl: Al/Alloy 5251-H22

DO NOT SCALE

01 Jan 2006

Sheet 1 of 1

Item: Chimney Brackets

Chk:

Drawn: KRS

Drg No:

BAAO30B

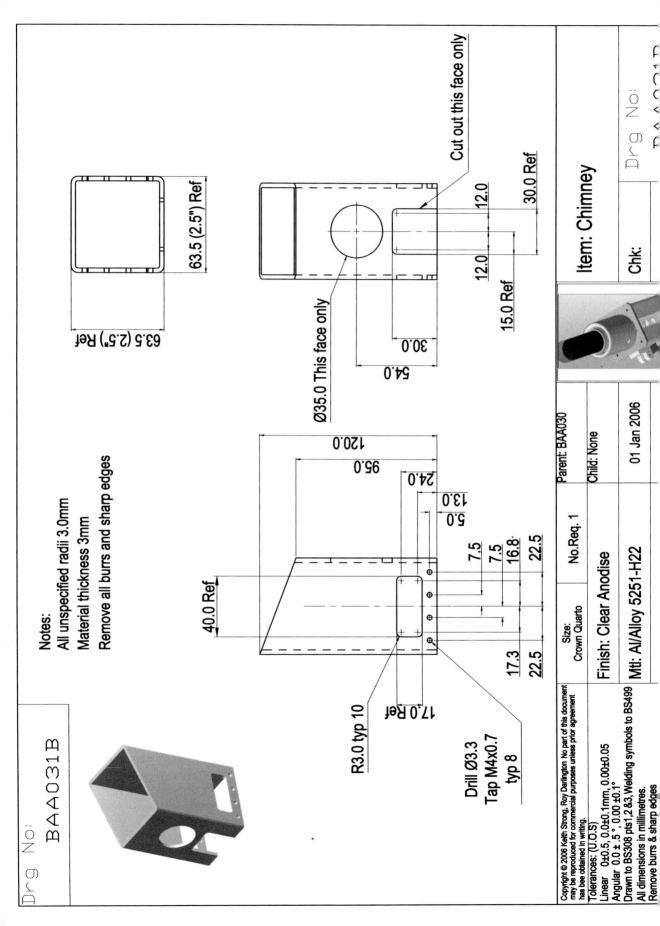

Notes:

All unspecified radii 3.0mm

Material thickness 3mm

Remove all burrs and sharp edges

63.5 (2.5") Ref

63.5 (2.5") Ref

Cut out this face only

12.0

30.0 Ref

12.0

15.0 Ref

Ø35.0 This face only

30.0

54.0

120.0

95.0

24.0

13.0

5.0

7.5

7.5

16.8

22.5

40.0 Ref

17.3

22.5

R3.0 typ 10

17.0 Ref

Drill Ø3.3
Tap M4x0.7
typ 8

Item: Chimney

Chk:

Parent: BAA030

Child: None

01 Jan 2006

Size:
Crown Quarto

No.Req. 1

Finish: Clear Anodise

Mtl: Al/Alloy 5251-H22

Copyright © 2006 Keith Strong, Roy Darlington No part of this document may be reproduced for commercial purposes unless prior agreement has bee obtained in writing.

Tolerances: (U.O.S)
Linear 0±0.5, 0.0±0.1mm, 0.00±0.05
Angular 0.0 ± .5°, 0.00 ±0.1°
Drawn to BS308 pts 1,2 &3, Welding symbols to BS499
All dimensions in millimetres.
Remove burrs & sharp edges

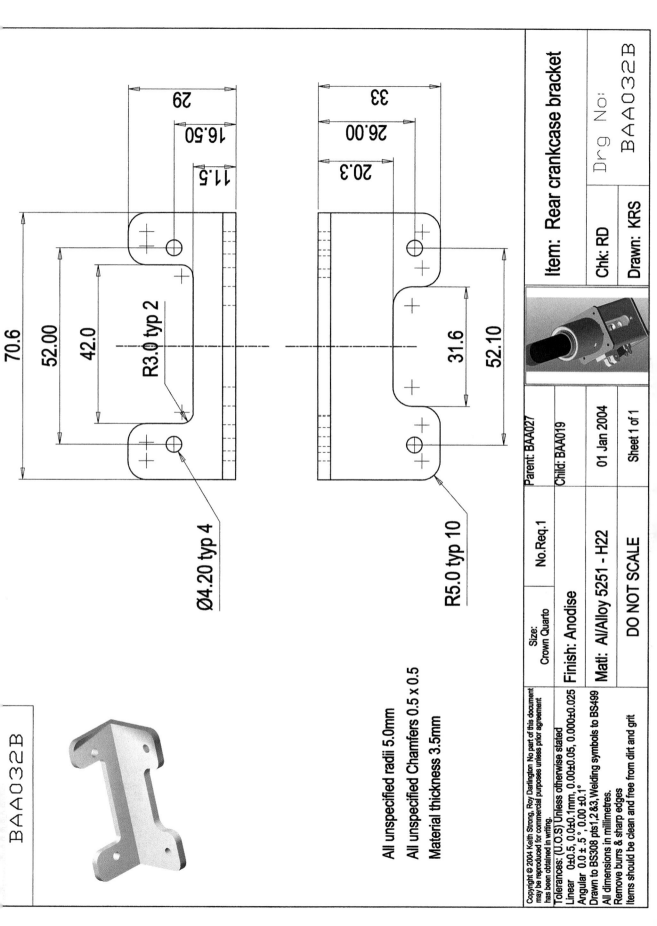

BAA032B

70.6
52.00
42.0
R3.0 typ 2
29
16.50
11.5
Ø4.20 typ 4

33
26.00
20.3
31.6
52.10
R5.0 typ 10

All unspecified radii 5.0mm
All unspecified Chamfers 0.5 x 0.5
Material thickness 3.5mm

Item: Rear crankcase bracket

Drg No: BAA032B

Chk: RD

Drawn: KRS

Parent: BAA027

No.Req.1

Child: BAA019

Size: Crown Quarto

Finish: Anodise

Matl: Al/Alloy 5251 - H22

01 Jan 2004

DO NOT SCALE

Sheet 1 of 1

Copyright © 2004 Keith Strong, Roy Darlington No part of this document may be reproduced for commercial purposes unless prior agreement has been obtained in writing.
Tolerances: (U.O.S) Unless otherwise stated
Linear 0±0.5, 0.0±0.1mm, 0.00±0.05, 0.000±0.025
Angular 0.0 ± .5°, 0.00 ±0.1°
Drawn to BS308 pts1,2 &3, Welding symbols to BS499
All dimensions in millimetres.
Remove burrs & sharp edges
Items should be clean and free from dirt and grit

Drg No: BAA033A

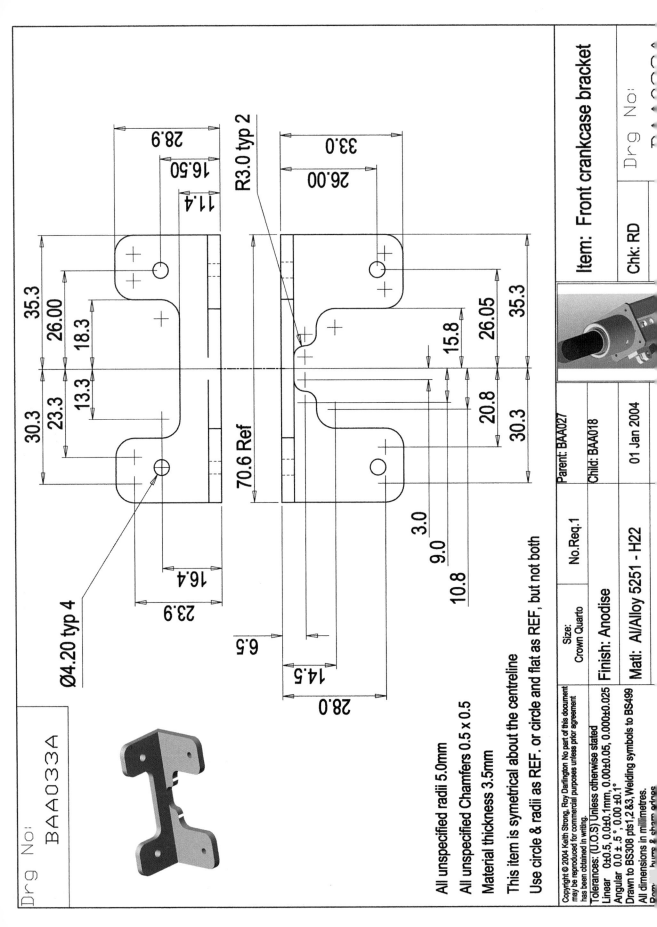

Ø4.20 typ 4

R3.0 typ 2

28.9
16.50
11.4
35.3
26.00
18.3
13.3
23.3
30.3

33.0
26.00
15.8
26.05
20.8
35.3
30.3

70.6 Ref

3.0
9.0
10.8

16.4
23.9

6.5
14.5
28.0

All unspecified radii 5.0mm

All unspecified Chamfers 0.5 x 0.5

Material thickness 3.5mm

This item is symetrical about the centreline

Use circle & radii as REF. or circle and flat as REF, but not both

Copyright © 2004 Keith Strong, Roy, Darlington No part of this document may be reproduced for commercial purposes unless prior agreement has been obtained in writing.

Tolerances: (U.O.S) Unless otherwise stated
Linear   0±0.5, 0.0±0.1mm, 0.00±0.05, 0.000±0.025
Angular 0.0 ± .5˚ , 0.00 ±0.1˚
Drawn to BS308 pts1,2 &3, Welding symbols to BS499
All dimensions in millimetres.
Rem... burrs & sharp edges

Size: Crown Quarto	No.Req.1	Parent: BAA027
Finish: Anodise		Child: BAA018
Matl: Al/Alloy 5251 - H22		01 Jan 2004

Item: Front crankcase bracket

Chk: RD

Drg No:

BAA034A

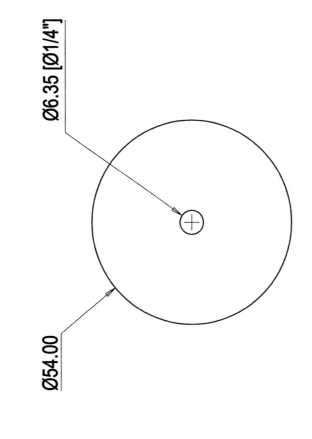

Ø6.35 [Ø1/4"]

Ø54.00

All unspecified radii 5.0mm

All unspecified Chamfers 0.5 x 0.5

Material thickness 4.0mm

Copyright © 2004 Keith Strong, Roy Darlington No part of this document may be reproduced for commercial purposes unless prior agreement has been obtained in writing.

	Size: Crown Quarto	No.Req.1	Parent: BAA19	Item:Cranckcase inspection window
Tolerances: (U.O.S) Unless otherwise stated Linear 0±0.5, 0.0±0.1mm, 0.00±0.05, 0.000±0.025 Angular 0.0 ± .5° , 0.00 ±0.1°	Finish: n/a		Child: None	
Drawn to BS308 pts1,2 &3, Welding symbols to BS499	Matl: Clear perspex		01 Jan 2004	Chk: RD
All dimensions in millimetres. Remove burrs & sharp edges Items should be clean and free from dirt and grit	DO NOT SCALE		Sheet 1 of 1	Drawn: KRS

Drg No: BAA034A

All unspecified Chamfers 0.5 x 0.5

Ø5.00
Ø9.00
14.20
M3X0.5
10 Deep
10.00
3.10

BAA041 x 1
DISPLACER BIG END

Ø5.00
Ø7.90
3.00

BAA037 x 3
ROCKER ARM BUSH

Ø5.00
Ø7.90
Ø9.40
8.00

BAA035 x 2
ROCKER PIVOT BUSH

Ø6.35
Ø9.00
3.00

BAA039 x 6
POWER PISTON CONROD BUSH X 4
ROCKER LINK BUSH X 2

Ø6.80
Ø3.90
Ø16.00
7.00
30.00

BAA047 x 1
POWER PISTON DISPLACER
CONROD BUSH

Ø14.00
Ø10.00
12.00
5.00

BAA053 x 1
DRIVE POST BUSH

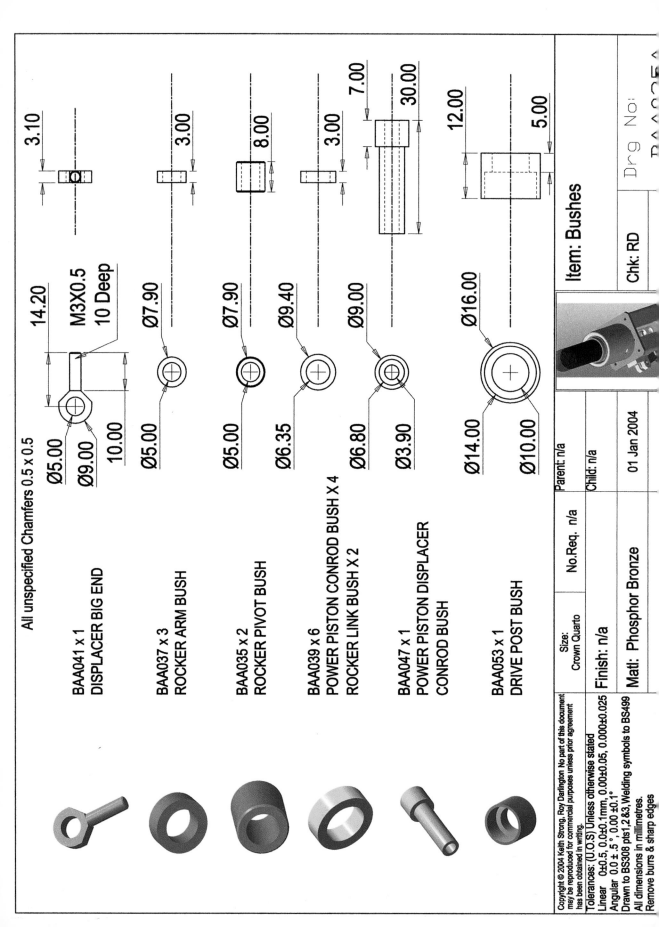

Item: Bushes

Chk: RD

Drg No:

Copyright © 2004 Keith Strong, Roy Darlington No part of this document may be reproduced for commercial purposes unless prior agreement has been obtained in writing.

Tolerances: (U.O.S) Unless otherwise stated
Linear   0±0.5, 0.0±0.1mm, 0.00±0.05, 0.000±0.025
Angular 0.0 ± 5 ° , 0.00 ±0.1°
Drawn to BS308 pts1,2 &3, Welding symbols to BS499
All dimensions in millimetres.
Remove burrs & sharp edges

Size:
Crown Quarto

Finish: n/a

Matl: Phosphor Bronze

No.Req.   n/a

Parent: n/a

Child: n/a

01 Jan 2004

BAA040A

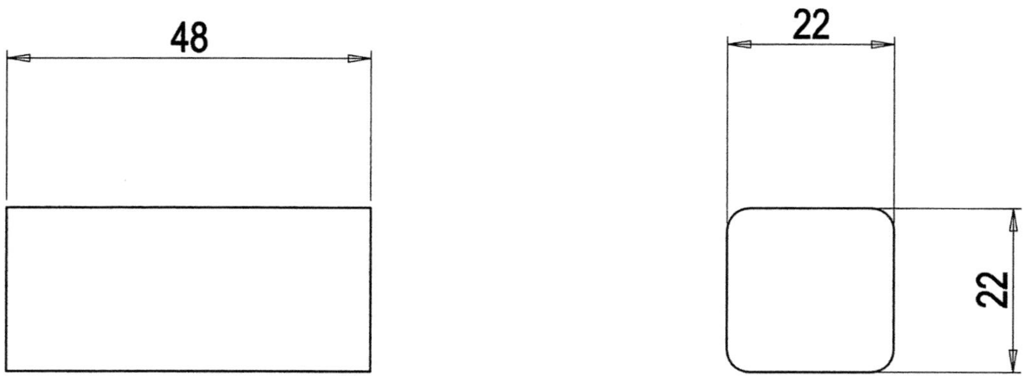

22

22

48

Copyright © 2004 Keith Strong, Roy Darlington No part of this document may be reproduced for commercial purposes unless prior agreement has been obtained in writing.

Tolerances: (U.O.S) Unless otherwise stated
Linear   0±0.5, 0.0±0.1mm, 0.00±0.05
Angular 0.0 ± .5˚, 0.00 ±0.1˚
Drawn to BS308 pts1,2 &3, Welding symbols to BS499
All dimensions in millimetres.
Remove burrs & sharp edges
Items should be clean and free from dirt and grit

Size:
Crown Quarto

No.Req. 1

Parent: BAA011

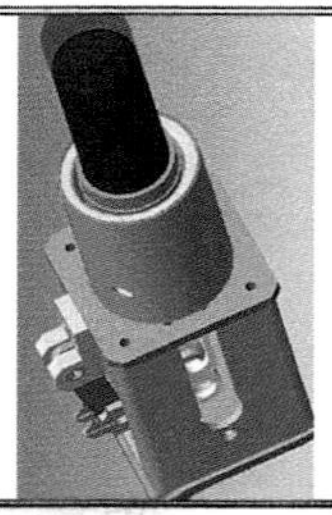

Finish: None

Child: None

Mtl: Stainless steel mesh

01 Jan 2004

DO NOT SCALE

Sheet 1 of 1

Item: Kettle defurrer
(Bought in part)

Chk: RD

Drg No:
BAA040A

Drawn: KRS

Ø4.90

Tap M3x0.5
Full length

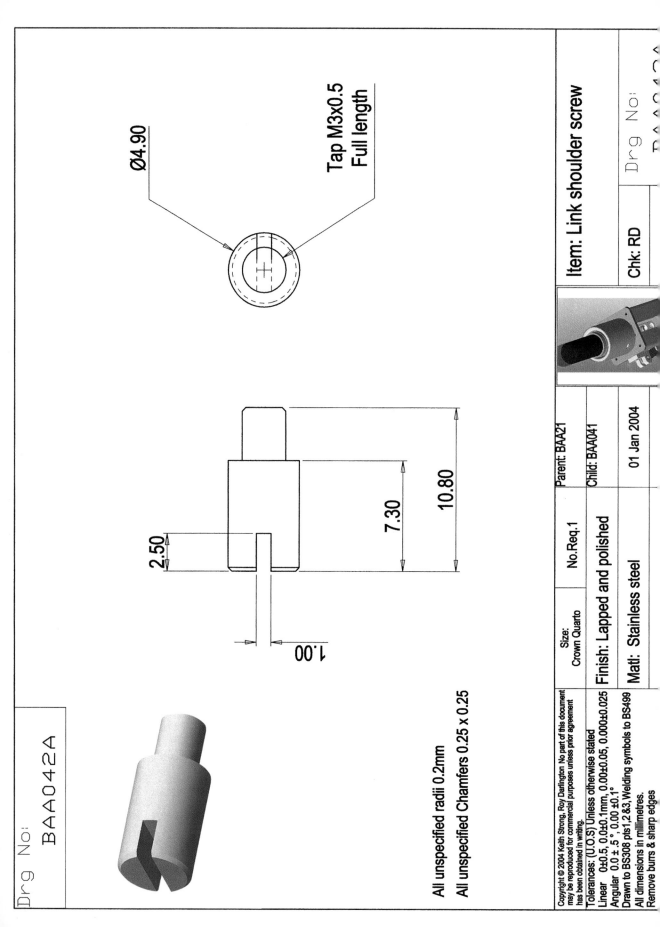

2.50

7.30

10.80

1.00

All unspecified radii 0.2mm
All unspecified Chamfers 0.25 x 0.25

Item: Link shoulder screw

Drg No:

Chk: RD

Copyright © 2004 Keith Strong, Roy Darlington No part of this document
may be reproduced for commercial purposes unless prior agreement
has been obtained in writing.

Tolerances: (U.O.S) Unless otherwise stated
Linear   0±0.5, 0.0±0.1mm, 0.00±0.05, 0.000±0.025
Angular 0.0 ± 5˚, 0.00 ±0.1˚
Drawn to BS308 pts1,2 &3, Welding symbols to BS499
All dimensions in millimetres.
Remove burrs & sharp edges

Size:
Crown Quarto

No.Req.1

Parent: BAA21

Child: BAA041

Finish: Lapped and polished

Matl: Stainless steel

01 Jan 2004

BAA045A

Ø29.2

Ø6.0

31.000

31.700

Drill Ø3.30
Tap M4x0.7

A

A

10

2.4

8

SECTION A-A

All unspecified radii 3.5mm

All unspecified Chamfers 0.25 x 0.25

Take careful note of the tolerances

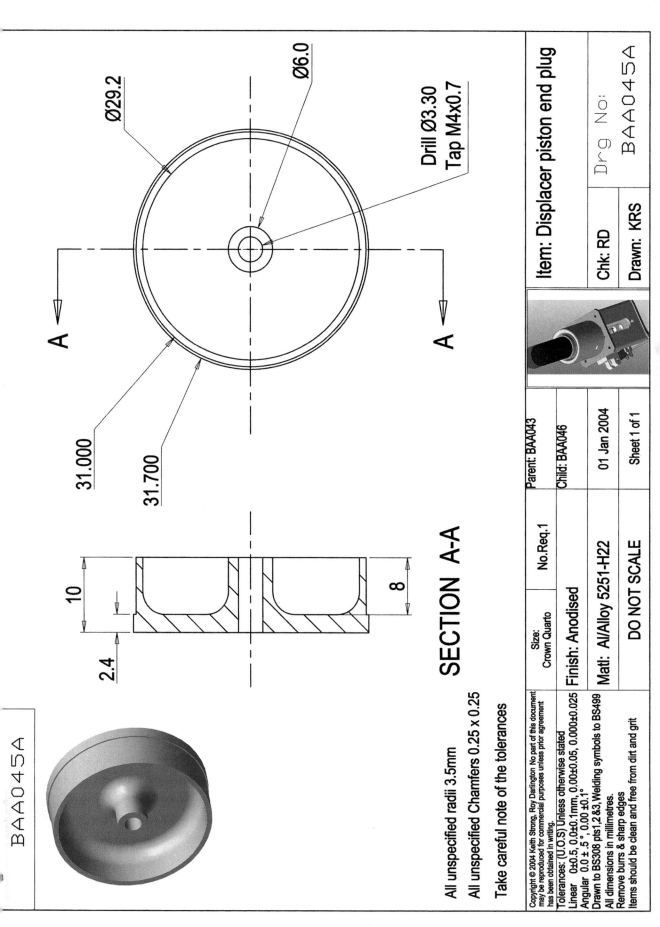

Item: Displacer piston end plug

Drg No:

BAA045A

Chk: RD

Drawn: KRS

Copyright © 2004 Keith Strong, Roy Darlington No part of this document may be reproduced for commercial purposes unless prior agreement has been obtained in writing.

Tolerances: (U.O.S) Unless otherwise stated
Linear   0±0.5, 0.0±0.1mm, 0.00±0.05, 0.000±0.025
Angular 0.0 ±.5 ° 0.00 ±0.1°
Drawn to BS308 pts1,2 &3, Welding symbols to BS499
All dimensions in millimetres.
Remove burrs & sharp edges
Items should be clean and free from dirt and grit

Size:
Crown Quarto

No.Req.1

Parent: BAA043

Finish: Anodised

Child: BAA046

Matl: Al/Alloy 5251-H22

01 Jan 2004

DO NOT SCALE

Sheet 1 of 1

All unspecified radii 2.5mm

All unspecified Chamfers 0.25 x 0.25

Take careful note of the tolerances

Ø31.70 $^{+0.000}_{-0.025}$

Ø31.20 $^{+0.000}_{-0.025}$

90.75

92.25

R3.00

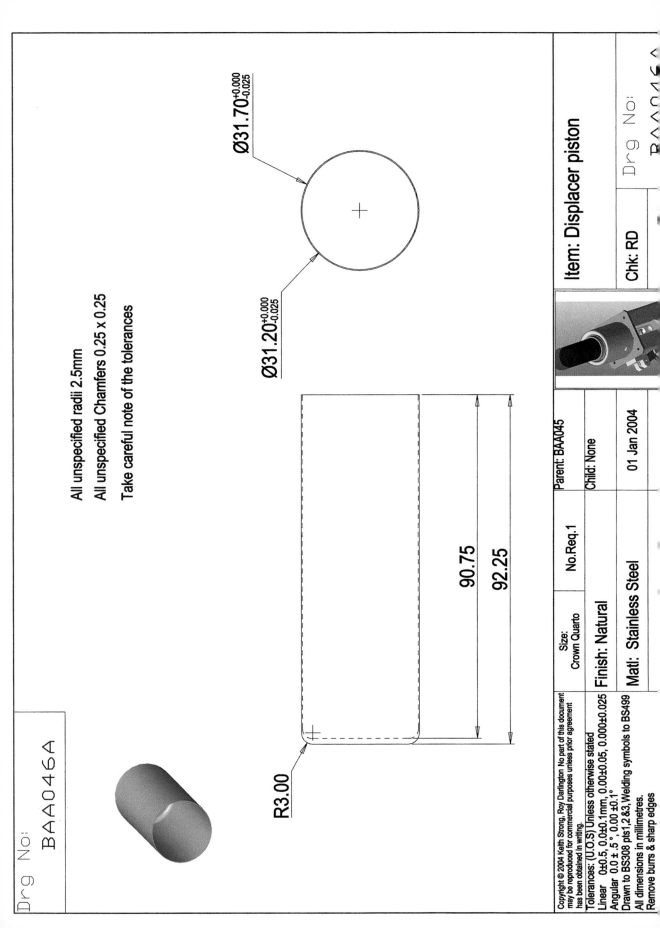

Item: Displacer piston

Drg No:

Chk: RD

BAA046A

Parent: BAA045

Child: None

No.Req.1

Size:
Crown Quarto

Finish: Natural

Matl: Stainless Steel

01 Jan 2004

Copyright © 2004 Keith Strong, Roy Darlington No part of this document
may be reproduced for commercial purposes unless prior agreement
has been obtained in writing.

Tolerances: (U.O.S) Unless otherwise stated
Linear   0±0.5, 0.0±0.1mm, 0.00±0.05, 0.000±0.025
Angular 0.0 ± .5 °, 0.00 ±0.1°
Drawn to BS308 pts1,2 &3, Welding symbols to BS499
All dimensions in millimetres.
Remove burrs & sharp edges

BAA048B

Chamfer 1.0 Typ 2

20.0

13.0

12.5

Ø10

Notes:
All unspecified chamfers 0.25mm

25.0

15.0

12.5

Drill Ø3.30
Tap M4 x 0.7

13.0

Dril Ø2.50
Tap M3 x 0.5

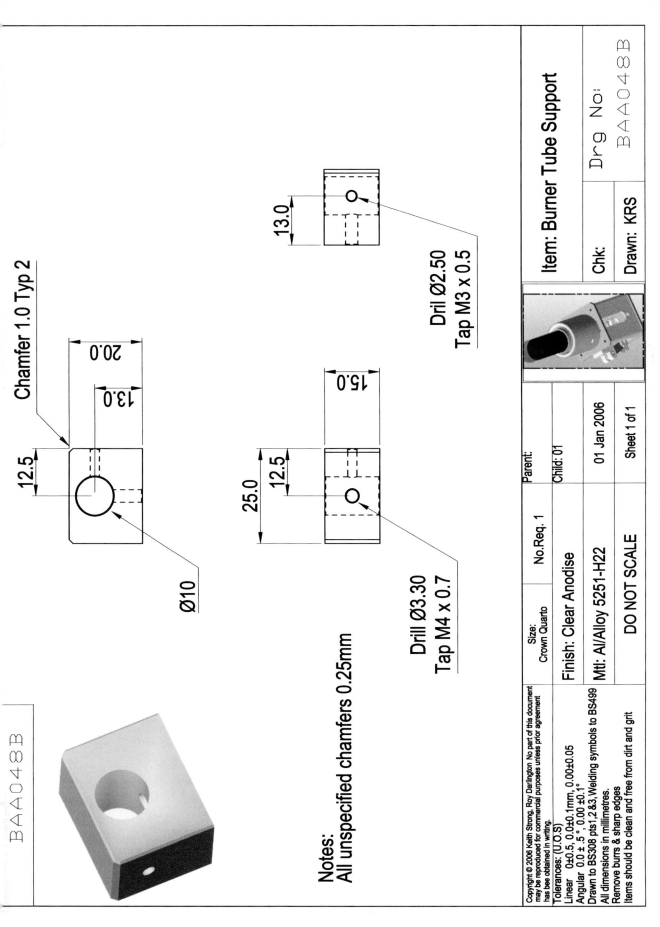

Copyright © 2006 Keith Strong, Roy Darlington No part of this document may be reproduced for commercial purposes unless prior agreement has bee obtained in writing.

Tolerances: (U.O.S)
Linear   0±0.5, 0.0±0.1mm, 0.00±0.05
Angular 0.0 ± .5°, 0.00 ±0.1°
Drawn to BS308 pts1,2 &3, Welding symbols to BS499
All dimensions in millimetres.
Remove burrs & sharp edges
Items should be clean and free from dirt and grit

		Parent:	
Size: Crown Quarto	No.Req. 1	Child: 01	Item: Burner Tube Support
Finish: Clear Anodise			
Mtl: Al/Alloy 5251-H22		01 Jan 2006	Chk:
DO NOT SCALE		Sheet 1 of 1	Drawn: KRS

Drg No:
BAA048B

Drg No:
BAA050A

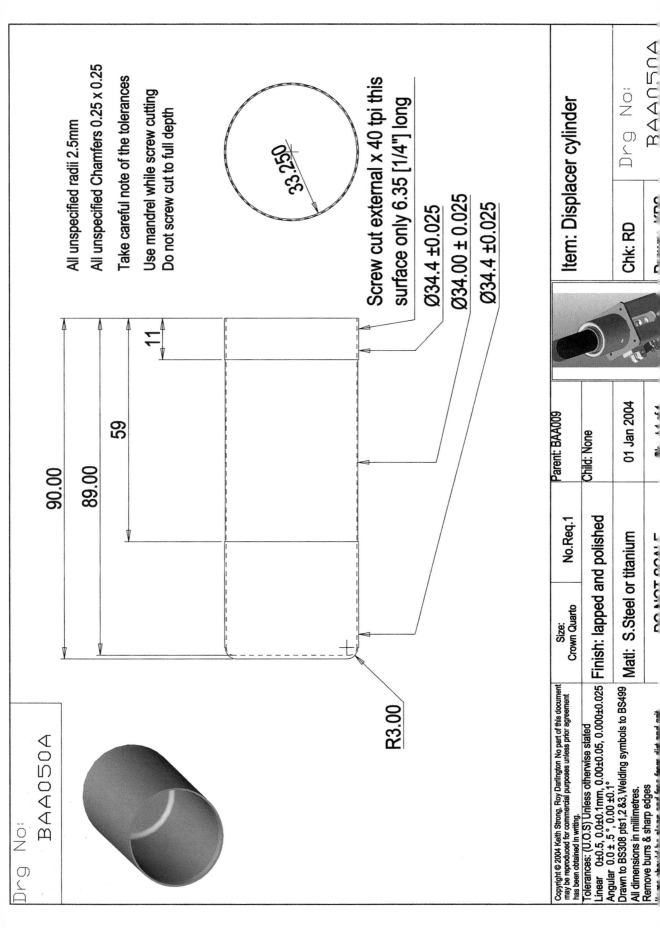

All unspecified radii 2.5mm
All unspecified Chamfers 0.25 x 0.25
Take careful note of the tolerances
Use mandrel while screw cutting
Do not screw cut to full depth

33.250

Screw cut external x 40 tpi this
surface only 6.35 [1/4"] long

Ø34.4 ±0.025
Ø34.00 ± 0.025
Ø34.4 ±0.025

90.00

89.00

59

11

R3.00

Copyright © 2004 Keith Strong, Roy Darlington No part of this document
may be reproduced for commercial purposes unless prior agreement
has been obtained in writing.

Tolerances: (U.O.S) Unless otherwise stated
Linear 0±0.5, 0.0±0.1mm, 0.00±0.05, 0.000±0.025
Angular 0.0 ± 5° , 0.00 ±0.1°
Drawn to BS308 pts1,2 &3, Welding symbols to BS499
All dimensions in millimetres.
Remove burrs & sharp edges

Size:
Crown Quarto

No.Req.1

Parent: BAA009

Child: None

Finish: lapped and polished

Matl: S.Steel or titanium

01 Jan 2004

Item: Displacer cylinder

Chk: RD

Drg No:
BAA050A

BAA051A

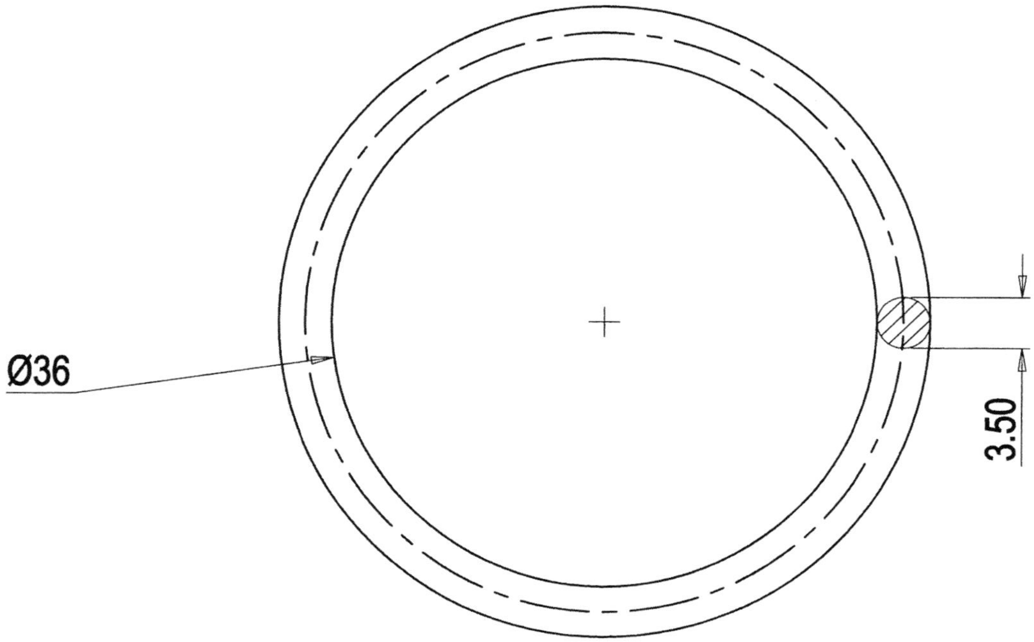

3.50

Ø36

Item: O-Ring

Drg No: BAA051A

Chk: RD

Drawn: KRS

Parent: BAA009

Child: BAA020

01 Jan 2004

Sheet 1 of 1

Size:
Crown Quarto

No.Req.2

Finish: N/A

Matl: Synthetic Rubber

DO NOT SCALE

Copyright © 2004 Keith Strong, Roy Darlington. No part of this document may be reproduced for commercial purposes unless prior agreement has been obtained in writing.
Tolerances: (U.O.S) Unless otherwise stated
Linear 0±0.5, 0.0±0.1mm, 0.00±0.05, 0.000±0.025
Angular 0.0 ± .5° , 0.00 ±0.1°
Drawn to BS308 pts1,2 &3, Welding symbols to BS499
All dimensions in millimetres.
Remove burrs & sharp edges
Items should be clean and free from dirt and grit

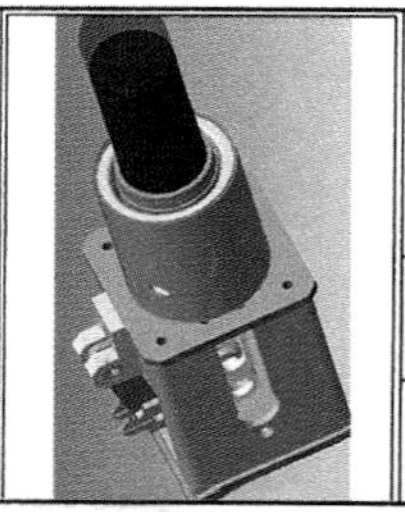

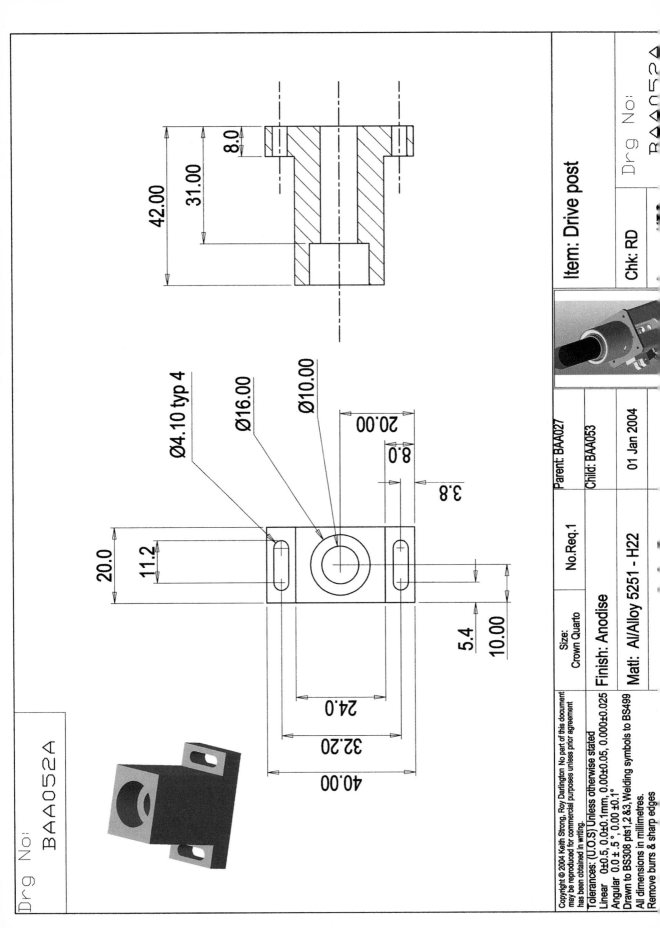

42.00

31.00

8.0

Ø4.10 typ 4

Ø16.00

Ø10.00

20.00

8.0

3.8

20.0

11.2

5.4

10.00

24.0

32.20

40.00

Item: Drive post

Parent: BAA027

No.Req.1

Size:
Crown Quarto

Finish: Anodise

Matl: Al/Alloy 5251 - H22

Child: BAA053

01 Jan 2004

Chk: RD

Drg No:

BAA052

Copyright © 2004 Keith Strong, Roy Darlington. No part of this document
may be reproduced for commercial purposes unless prior agreement
has been obtained in writing.

Tolerances: (U.O.S) Unless otherwise stated
Linear 0±0.5, 0.0±0.1mm, 0.00±0.05, 0.000±0.025
Angular 0.0 ± .5 ˚, 0.00 ±0.1°
Drawn to BS308 pts1,2 &3, Welding symbols to BS499
All dimensions in millimetres.
Remove burrs & sharp edges

BAA054A

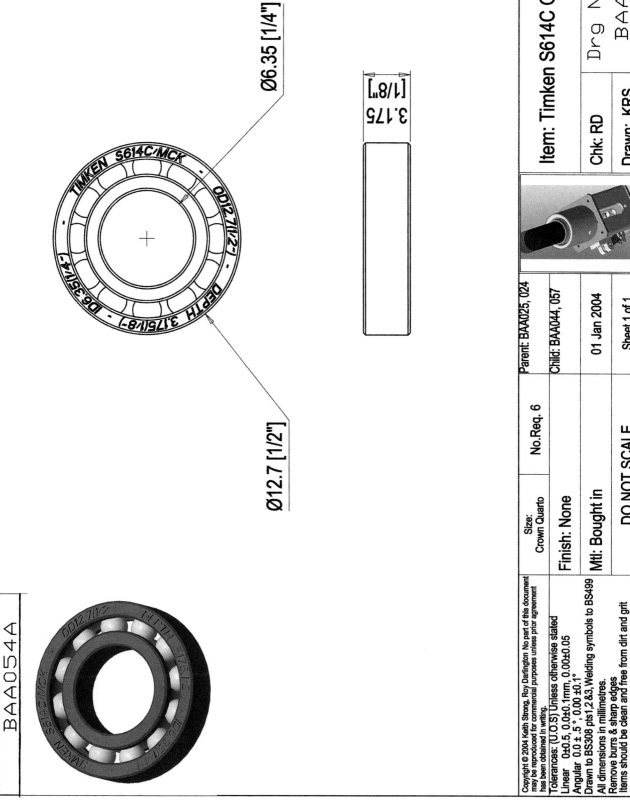

Ø6.35 [1/4"]

Ø12.7 [1/2"]

3.175 [1/8"]

Item: Timken S614C OR MCK

Drg No: BAA054A

Chk: RD

Drawn: KRS

Copyright © 2004 Keith Strong, Roy Darlington. No part of this document may be reproduced for commercial purposes unless prior agreement has been obtained in writing.

Tolerances: (U.O.S) Unless otherwise stated
Linear   0±0.5, 0.0±0.1mm, 0.00±0.05
Angular 0.0 ± .5° 0.00 ±0.1°
Drawn to BS308 pts1,2 &3, Welding symbols to BS499
All dimensions in millimetres.
Remove burrs & sharp edges
Items should be clean and free from dirt and grit

Size: Crown Quarto

No.Req. 6

Parent: BAA025, 024

Finish: None

Child: BAA044, 057

Mtl: Bought in

01 Jan 2004

DO NOT SCALE

Sheet 1 of 1

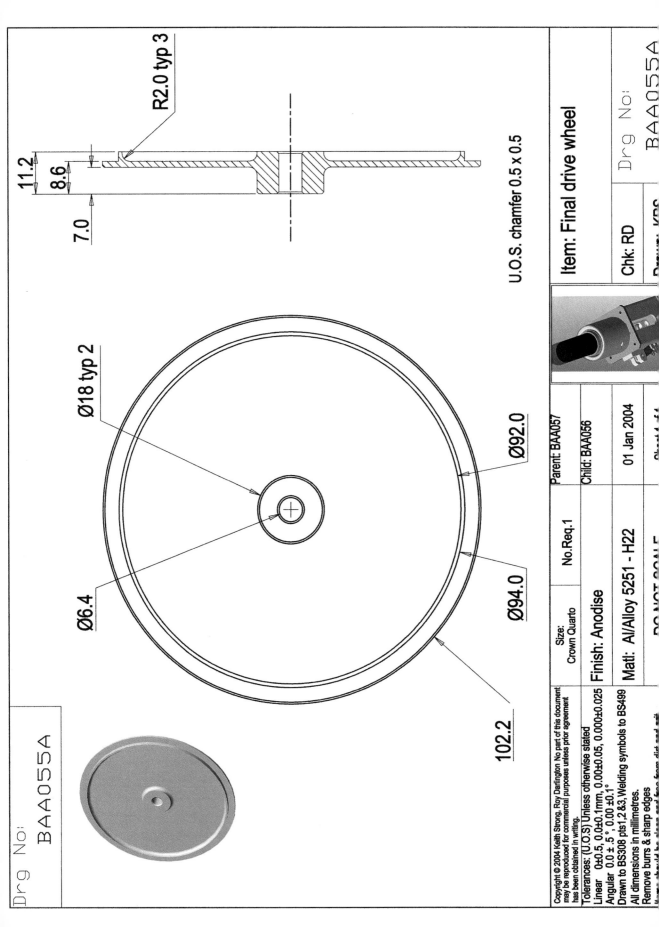

Drg No:
BAA055A

R2.0 typ 3

11.2

8.6

7.0

Ø18 typ 2

Ø6.4

Ø92.0

Ø94.0

102.2

U.O.S. chamfer 0.5 x 0.5

Item: Final drive wheel

Drg No:
BAA055A

Chk: RD

Copyright © 2004 Keith Strong, Roy Darlington No part of this document may be reproduced for commercial purposes unless prior agreement has been obtained in writing.

Tolerances: (U.O.S) Unless otherwise stated
Linear   0±0.5, 0.0±0.1mm, 0.00±0.05, 0.000±0.025
Angular 0.0 ± .5° , 0.00 ±0.1°
Drawn to BS308 pts1,2 &3, Welding symbols to BS499
All dimensions in millimetres.
Remove burrs & sharp edges

Parent: BAA057

Child: BAA056

Size:
Crown Quarto

No.Req.1

01 Jan 2004

Finish: Anodise

Matl: Al/Alloy 5251 - H22

BAA056A

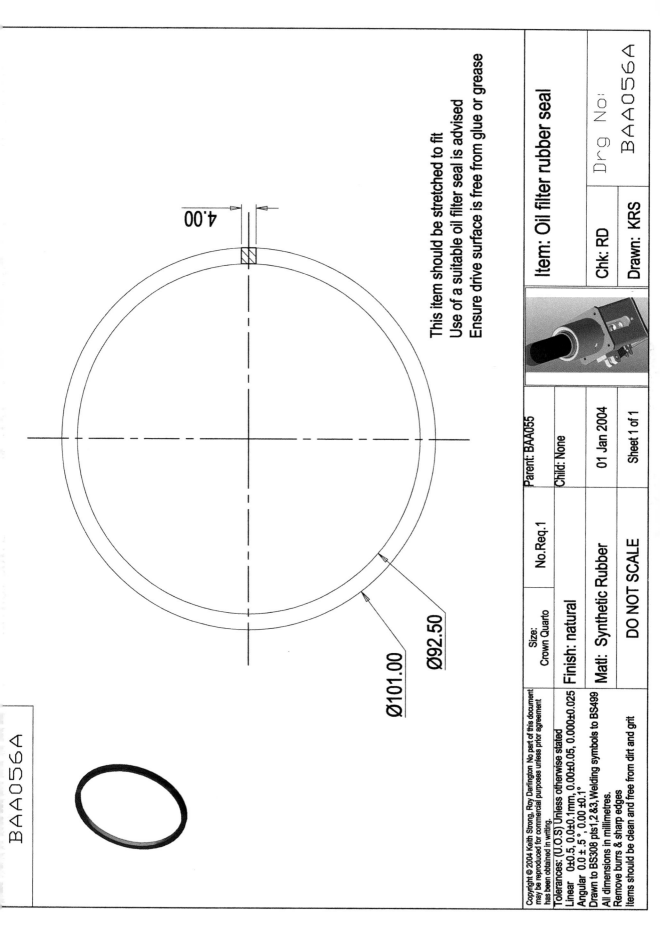

4.00

Ø101.00

Ø92.50

This item should be stretched to fit
Use of a suitable oil filter seal is advised
Ensure drive surface is free from glue or grease

Copyright © 2004 Keith Strong, Roy Darlington No part of this document may be reproduced for commercial purposes unless prior agreement has been obtained in writing.

Tolerances: (U.O.S) Unless otherwise stated
Linear   0±0.5, 0.0±0.1mm, 0.00±0.05, 0.000±0.025
Angular 0.0 ± .5°, 0.00 ±0.1°
Drawn to BS308 pts1,2 &3, Welding symbols to BS499
All dimensions in millimetres.
Remove burrs & sharp edges
Items should be clean and free from dirt and grit

Size: Crown Quarto	No.Req.1	Parent: BAA055
Finish: natural		Child: None
Matl: Synthetic Rubber		01 Jan 2004
DO NOT SCALE		Sheet 1 of 1

Item: Oil filter rubber seal

Chk: RD

Drawn: KRS

Drg No: BAA056A

Glossary

Ackermann A system in which the wheels that steer attach to hubs and the angle of turn is controlled by a steering arm. The angles of the steering arms are relative to the centre line of the vehicle and are positioned on lines which typically cross the centre line of the vehicle at the centre of the rear axle or differential. This layout means that the front wheels are able to steer at an appropriate angle relative to their individual arc for any given turn. This optimizes the turn angle and minimizes the rolling friction and slip angle for the tyres. This means better economy and more usable power.

Annealing A heat treatment technique that is used to relieve the work hardening of metal by softening it. This is achieved by heating it to a specific temperature relative to the metal, soaking (keeping it at that temperature) for a further time period and then cooling it to a further specified temperature inside an oven or immersed in a powder. This cools the component very slowly and relieves any internal stresses present in the metal. In general, the shorter the cooling period the more brittle the metal will be. Most metals can have considerably different properties just by altering the heat treatment applied.

Anodizing To coat metal with a protective coating using electrolysis; usually used with aluminium, it gives a glass-like coating of aluminium oxide that may be coloured using dyes.

Ball race Ball bearings held together with a cage usually between an inner and outer hardened steel ring. The rings will often have a groove or channel to captivate the ball bearings.

BDC Bottom Dead Centre. This is the point that a piston reaches at the lowest part of its stroke.

Beam The width of a boat. Also a component used in some Stirling engines as a kind of rocker that reciprocates between the power and displacer pistons.

Big end The end of the con-rod that attaches to the crank, usually larger than the small end.

Bleed valve A device that allows pressure relief. A simple bleed valve can be a ball bearing pushed against a lip by a compression spring. The pressure relief setting can be varied by more or less compression of the spring.

Brazing A form of joining metals (typically steel) with heat utilizing higher temperatures than with silver-soldering. A brazing torch is used and the filler metal is brass alloy. This creates a stronger joint than in soldering or silver-soldering and is the strongest method of creating a joint without severely altering the structure of the metal (as in welding).

Brittleness Brittle materials will break with very little twisting or bending. Hard materials are generally brittle.

Broken This is a term used to describe the act of and the result of deburring. This leaves a small chamfer on edges, corners or holes and is primarily specified where parts may be handled. It eliminates injury from razor sharp edges and pieces of swarf that may be protruding from a component after machining.

Case hardening The metal is heated to a specific temperature while in a solid, liquid or gas, which in the case of steel deposits carbon on to the surface. Upon rapid cooling the surface becomes hard. Types of case hardening that are common are nitriding and carburizing.

Clearance hole A clearance hole is used to allow a bolt to pass through it with room to spare. To find the relevant clearance hole for a particular bolt use a look-up chart like 'Zeus' precision charts. For example: A clearance hole for an M3 bolt is 3.1mm. (Also *see* Tapping hole).

Clevis A kind of universal coupling that only pivots in one plane. An alternative method of connecting the gudgeon pin to a little end. It can also be used to join two shafts together where the motion changes direction. Often a fork-like part, usually with two tines between which a pivot or a bearing will connect to a shaft.

Collet A clamping device used on a shaft.

Compressive strength The measurement that allows comparison of the ability of a material to withstand a pressure or weight.

Conductivity The rate of electrical or heat transfer in a material.

Con-rod Short for connecting rod. It has a big end connected to a crank and a little end connected to a piston.

Counterbore A hole that is bored where the inside base of the hole is flat. It can be either a blind hole or an addition to an existing hole, like a spot face with depth.

Countersink A chamfer cut on to a hole to accommodate a flat-headed screw. Different angles of cut are used and the countersink cutter should be chosen to suit the angle of the intended screws. Countersink instructions can be specified with either an angle of the cutter or the base diameter dimension followed by the top diameter dimension: i.e. Dia. 3.0 Cs'k 90deg. or Dia. 3.0 Cs'k to Dia.6.5.

Crank disc Usually a round disc with an offset pin or a hole that serves as an eccentric. It enables rotary motion to be converted into reciprocating motion.

Crankshaft This allows rotary motion to be turned into reciprocating motion and is an alternative to a crank disc. It usually has con-rods attached and is also coupled to the flywheel. The parts of the crankshaft that accommodate the big end bearings are called journals.

Dead space In a Stirling engine when the power piston reaches TDC there is a space that contains the air or working gas. This is known as the dead space. This term is also applied to the extended air gap such as connecting tubes or ports coming from the top of the power piston until it fills the volume of the displacer cylinder.

Diaphragm A flexible barrier that has a rod attached to a flywheel and replaces the normal style of power piston. It is usually made from some form of synthetic rubber.

Double acting A method of utilizing the underside volume of a piston for feeding the top volume of another. These will typically be arranged in a group of four.

DTI Dial Test Indicator. An instrument for accurately measuring the location of a surface.

Ductility Often confused with malleability, ductile materials tend to be weak and soft and can be drawn into thin sections like wire.

Ductility is normally used to mean both malleable and ductile.

Dynamometer A machine that is used to measure the performance of an engine.

Eccentric A description that refers to a hole or part that rotates not using its centre point. Typically describes cranks or cams.

Elasticity This is the property of a material to return to its original shape after deformation.

Face/Spot face The process of making a flat or smooth surface on a piece of material. Spot face is generally circular and is sized and applied where a pivot or a nut may be required to turn or bear against.

Fine feed A term which describes the rate of travel of a cutting tool across or into the work. It is also used as a description for the gearing on a lathe that allows this to happen.

Flux A cleaning agent that cleans the metal chemically. It allows maximum adhesion of molten metals used in hot joining processes such as silver-soldering, brazing and soldering.

Focal point *See* Parabolic reflector.

Gudgeon pin This is used in a piston and connects the piston to the small end of the con-rod. It will typically be made from a hardened material like silver steel and allows the piston and con-rod to have relative angular movement.

Hardening This may be applied to a metal by heating it to a specific temperature and then quenching it in oil, water or brine. There will be an increase in the brittleness of the metal.

Hardness Resistance to damage or wear from sharp points or bearing objects such as shafts and seals.

Heat sink This is usually a finned component that is designed to draw heat away from another component.

Honing A honing tool is usually a three-lobed tool that has a whetstone on each of its sprung-loaded lobes. It is used for grinding or polishing cylinders primarily to remove tool marks. This can also be applied to the action of abrading just prior to lapping or polishing, and after machining.

Hot cap The end of a displacer cylinder where the flame or heat is applied to the cylinder. Occasionally the whole of the displacer is incorrectly referred to as the hot cap.

HSS High Speed Steel. This is steel with a high carbon content. It is very hard and very brittle, and broken drills will testify to this.

IC Internal Combustion or spark ignition engine as opposed to a CI engine which is short for Compression Ignition (diesel). IC is also an abbreviation used to mean an Integrated Circuit (silicon chip).

Journal *See* Crankshaft

Knurled A decorative gripping surface of straight-line or diagonal relief made by uniformly serrated rollers. Using a lathe it is applied to a shaft with these rollers.

Lapping A type of polishing or grinding that is used as a final surface preparation before any polishing that may be required is done; usually it is used to remove light tool marks. Simple tools can be made using a dowel rod that has been split, and is held open or expanded with a wood screw.

LED Light Emitting Diode. An electronic component available in a variety of colours that allows current to flow in a circuit in one direction and not the other. Typically it will be used in a 5V or 3.3V circuit. It will require an additional resistance of some description to stop it burning out.

Linishing A method of creating a uniform surface using a linishing machine. For use on flat surfaces prior to polishing. It quickly removes tool marks. It is effectively a bench- or floor-mounted belt sander.

Lip seal A thin seal that has a very fine feather edge that presses harder against a cylinder wall the more pressure that is applied to it.

Little end The end of the con-rod that attaches to the piston via the gudgeon pin. It is normally, but not always, smaller than the big end.

Low Δ T Low Delta Temperature. This terminology is used to name a particular derivative of engine that uses the Stirling principle. This type of engine requires only a very small difference in temperature between the two ends of the displacer cylinder to operate. The Greek delta Δ symbol is often used to describe a change or a rate of change of something, in this case temperature.

Machinability A measure of the ease of a material to be machined.

Mandrel A buck that is used to hold an item that is to be machined, often used where the component would get damaged from the clamping forces of a lathe chuck.

Malleable A description of material that bends without cracking.

Micron 1,000,000th of a metre or 1,000th of a millimetre. Written (1μm) or 0.001mm. Equivalent to 0.0000393701in.

Modulus of elasticity The amount of deformation a material will take for a certain amount of stress. Also called Young's Modulus after Thomas Young.

Moment of inertia The resistance of an object to rotation.

Normalizing This is similar to annealing but is done at a higher temperature and is kept at that temperature to ensure uniform temperature and then cooled in air. Normalized parts are stronger, harder and tougher than with annealing or tempering.

OD/ID Outside Diameter/Inside Diameter.

Oilite A type of plain bearing material that is sintered and capable of retaining oil. It is typically used for low-speed shafts, hinges and pivots. It contains graphite with very fine grains of phosphor bronze that are formed under great pressure.

Otto-cycle Otherwise known as a spark ignition engine (SI engine). This is what most people would consider to be a petrol engine.

Parabolic reflector This describes the shape of a reflective surface from which the heat or light that is reflected focuses on one central point. This is known as the focal point.

Parting tool A specific thin lathe tool that has a cutting edge that is square to the work and is usually used to cut a machined component from a bar that is held in the chuck of a lathe. It is useful for machining cooling fins.

PCD Pitch Circle Diameter. This is the diameter that the centre of a circle or hole is placed on. Usually the pitch is determined with equally spaced holes circumferentially on the specified diameter.

Peening The act of hardening the edge or surface of a metal using impact techniques. Holes can be peened by hardening the edges and surfaces can be peened in volume using shot peening. Peening hammers are used for decorative as well as work-hardening purposes on soft metals such as copper. This is sometimes called planishing.

Polishing The mechanical method of smoothing a metal surface usually achieved with a polishing wheel. Various finishes may be obtained with a wide variety of abrasive compounds applied to a polishing wheel.

PTFE Polytetrafluoroethylene. A very slippery type of plastic that is commonly used for bearing surfaces. Also known as Teflon or Fluon.

RPM Revolutions Per Minute.

Shear strength The measurement of the strength of a material to resist fracturing due to opposing forces such as a con-rod pulling on the pin of a clevis.

Silver-soldering The joining of two parts using an alloy called silver solder which melts between 650 and 875°C and is used for making strong yet moderately ductile joints that resist corrosion and have good heat-transfer properties.

SMEE Society of Model and Experimental Engineers (*see* 'Recommended Societies').

Soldering The joining together of two metal parts using an alloy of zinc, tin and lead. This technique is at a lower temperature than brazing or silver-soldering and is suitable for low-strength joints.

Tapping hole This is a specification applied to holes which gives an indication that a hole of a nominal size should be drilled undersize in order to allow for the cutting of a thread, for example for M3 you would expect to drill to a diameter of 2.5mm. This will vary depending on which standard is being followed and which pitch is being used on the thread. The exact specifications can be checked in look-up tables according to a calculation using the thread diameter and the pitch size. It should allow enough material to accommodate the appropriate pitch of the expected thread.

TDC Top Dead Centre. This is the point that a piston reaches at the highest part of its stroke.

Tempering Metals are tempered to reduce internal stresses and brittleness by heating the metals to a specific temperature and then allowing the metal to cool in air.

Tensile strength A measurement of the property of a material to withstand being pulled apart in a straight line. It is written in this book as Newtons per millimetre squared (N/mm sq.). It is also written as MPa which is mega Pascals and is identical to N/mm sq. It is also written as PSI or pounds per square inch, with a suitable factor applied.

Thermodynamics A term which describes the rate of change and the transfer of heat and therefore work, in an object, ultimately converting one form of energy into another.

TMG Thermo Mechanical Generator. A type of generator that produces electricity using the Stirling principle.

Topslide lead screw A threaded shaft on the lathe that controls the motion of the cross-feed carriage.

Toughness Resistance to fracture with the added benefit of resisting further failure once cracks have appeared.

Universal coupling/joint This is used to transmit rotary shaft power through an angle. Various types include fabric, Hooke, rubber trunnions, CV (Constant Velocity) joints and Rzeppa.

Useful Information

REFERENCES

BS 308 parts 1, 2 and 3 'Engineering drawing practice. Recommendations for general principles, 1993. Recommendations for dimensioning and tolerancing of size, 1985. Recommendations for geometrical tolerancing, 1990.' (British Standards Institution)
Superseded by:
BS 8888 'Technical product documentation (TPD). Specification for defining, specifying and graphically representing products.' (British Standards Institution 2002)

BS 1449 'Steel plate, sheet and strip. Carbon and carbon manganese plate, sheet and strip. Specification for hot rolled narrow strip supplied in a range of conditions for heat treatment and general engineering purposes.' (British Standards Institution 1991)

BS 4500 'ISO limits and fits' (British Standards Institute 1988)

Darlington, R. 'Noddy. A hot air engine for everybody', *Model Engineer* (Vol. 177, No. 4,027, 18 October 1996)

Jones, J. 'Low Temperature Differential Hot Air Engine' *Model Engineer* (Vol. 175, No. 4,001, 15 September 1995 and No. 4,003, 20 October 1995)

RECOMMENDED READING

Bedford, J. R. *Metalcraft Theory and Practice* (John Murray (Publishers) Ltd, 1972)
Gordon, J. E. *The New Science of Strong Materials* (Penguin Books, 1973)
Hibbeler, R. C. *Mechanics of Materials* (Pearson Education, 2003)
Hillier V. A.W. and Pittuck E. *Fundamentals of Motor Vehicle Technology* (Hutchinson Education, 1986)
Kolin, I. *The Evolution of the Heat Engine* (Moriya Press, 1998)
Memin, R. *Stirling. 4 Moteurs à Air Chaud* (Confrérie des Amateurs de Vapeur Vive, 2000)
Rizzo, J. G. *Modelling Stirling and Hot Air Engines* (Patrick Stephens, 1985)
Rizzo, J. G. *The Stirling Engine Manual* (Camden Miniature Steam Services, 1997)
Rizzo, J. G. *The Stirling Engine Manual,* Vol. II (Camden Miniature Steam Services, 1999)
Rogers, G. F. C. and Mayhew, Y. R. *Engineering Thermodynamics Work and Heat Transfer* (Longman, 1980)
Ross, A. *Stirling Cycle Engines* (Solar Engines, 1981)
Sarafin, T. P. and Wiley, J. L. *Spacecraft Structures and Mechanisms from Concept to Launch* (Microcosm, 1995)
Sellers, J. J. *Understanding Space An Introduction to Astronautics* (McGraw-Hill, 1994)
Senft, J. R. *An Introduction to Stirling Engines* (Moriya Press, 1993)
Senft, J. R. *Ringbom Stirling Engines* (Oxford University Press, 1993)
Senft, J. R. *An Introduction to Low Temperature Differential Stirling Engines* (Moriya Press, 1996)
Senft, J. R. *Miniature Ringbom Engines* (Moriya Press, 2000)
Sier, R. *A History of Hot Air and Caloric Engines* (Argus Books, 1987)
Sier, R. *Rev. Robert Stirling* (L. A. Mair, 1995)
Sier, R. *Hot Air Caloric and Stirling Engines,* Vol. I (Lindsay Publications, 1999)

Smith, C. *Engineer To Win* (Redwood Burn, 1988)

Society of Automotive Engineers *Automotive Handbook*, Third Edition (Robert Bentley, 1993)

Warbrooke, T. *Building Stirling One* (Camden Miniature Steam Services, 2002)

RECOMMENDED SOCIETIES

The Society of Model and Experimental Engineers

Marshall House, 28 Wanless Road, London SE24 0HW

http://www.sm-ee.co.uk

A new magazine *The Model Engineer and Amateur Electrician* appeared in January 1898. This was followed later in that year by the establishment of the Society of Model and Experimental Engineers. The Society has its own headquarters containing a workshop and a library of over 2,500 books and periodicals, which is probably the largest single collection related to the hobby of model engineering. The Society has always encouraged experimental work in all its forms: this is reflected in the collection of models that continues to grow and provides a physical history of the hobby. The Society publishes a journal six times a year containing technical articles and reports of the Society's activities. Regular meetings are held at Marshall House with lectures, discussions and occasional auctions featuring as regular activities.

The Stirling Engine Society

P.O. Box 5909, Chelmsford, Essex CM1 2FG

http://www.stirlingengines.org.uk

Founded in 1997. Its aim is to develop, educate and promote understanding of Stirling cycle engines. *Stirling News*, a quarterly newsletter, is produced and distributed to members worldwide.

SUPPLIERS

Recommendation is from my own experience. Details also appear on the website that complements this book:

http://www.stirlingengineer.com

Plans of other engines similar to the one shown in this book may be obtained from this website where interest, requests and comments may also be left.

Arc Euro Trade

10 Archdale Street, Syston, Leicester LE7 1NA

Supplier of small mills, lathes and engineering products. They specialize in ball, roller and needle roller race bearings for Low Δ T through to steel rolling mills.

http://www.arceurotrade.com

Bruce Engineering Model Supplies/Polly Model Engineering Limited

Bridge Court, Bridge Street, Long Eaton, Nottingham NG10 4QQ

For all model engineering requirements including materials, castings, fittings, accessories, tools, books, stationary engines and machined loco kits.

http://www.pollymodelengineering.co.uk

Sterling Stirling

15 The Pill, Caldicot, Newport, Monmouthshire NP26 5JQ

Information and materials for Stirling engine enthusiasts.

http://www.stirlingengines.org.uk

The College Engineering Supply

2 Sandy Lane, Codsall, Wolverhampton WV8 1EJ

Supplier of an extensive range of metal, engineering castings and materials that are suitable for the construction of hot air engines.

http://www.collegeengineering.co.uk

Warren Machine Tools Ltd

Warco House, Fisher Lane, Chiddingfold, Surrey GU8 4TD

Supplier of machine tools and accessories ranging from small to heavy duty.

http://www.warco.co.uk

Acknowledgments

With grateful thanks to:

Graham Darlington for his help and support over many years and with regard to this book, his advice on photography, lending photographic equipment and providing some of the photographs for this book.

Katie Strong for her enormous and invaluable contribution, listening to hours of tape deciphering my words into print.

James Rizzo, Geoff Barlett, Julian Wood, Bob Sier, Ken Boak, John Wilkinson and many other close friends including those in the Society of Model and Experimental Engineers, the Stirling Engine Society, Worthing and District Society of Model Engineers, Stamford Model Engineering Society and all the other model engineering societies and clubs that I visit regularly whose members have helped and encouraged me throughout the years. I hope I will be forgiven for not naming you all individually.

East Ayrshire Arts and Museums for permission to reproduce the photograph of Robert Stirling. East Ayrshire Arts and Museums manages: the Dick Institute, Kilmarnock; Dean Castle, Kilmarnock; the Palace Theatre and Grand Hall, Kilmarnock; the Baird Institute, Cumnock; Burns House Museum, Mauchline; and the Doon Valley Museum, Dalmellington. They have large and varied collections reflecting the natural, industrial, artistic and social heritage of south-west Scotland and beyond, and a lively mixture of exhibitions, events and performances. More information can be found at www.east-ayrshire.gov.uk

Tangmere Aviation Museum for their help in our research on Merlin engines and Lancasters.

West Sussex Fire and Rescue Service for advice on fire prevention and the safe use of flammable liquids.

Thanks must go to Yasmin Bathard, a mainstay of Roy's exhibition team, without whom Roy's continued presence at numerous shows and exhibitions would become less frequent, and the poorer for it.

Alistair McKitterick for assistance with the proof-reading.

Edward Proffitt for some of the initial CAD work.

We have been unable to trace the engineers who designed and made the London Marble and American Marble engines. If they wish to be added to future reprints, please contact the author, co-author or publisher.

The diagrams and most of the photographs in this book were taken or produced by Keith Strong. Where photographs were taken by other people they have been credited alongside. We were unable to trace the people who took two of the photographs of Roy Darlington in the RAF. If they wish to be added to future reprints, please contact the author, co-author or publisher.

APPENDIX I – MATERIAL DESCRIPTION AND USAGE

Aluminium

The ISO designation for aluminium is usually a four-digit number followed by a descriptor that designates the heat treatment. The various alloys available start with a number from 1 to 8 inclusive.

1 series (e.g. 1200 – almost pure aluminium) have good corrosion resistance, thermal and electrical conductivity, good workability but have low mechanical properties. The strength of this series can be marginally increased through heat treatment or work hardening.

2 series (e.g. 2024, 2014 – copper is the principal alloy) are widely used in some temper states, have very good yield properties and high strength. Other aluminium alloys have better corrosion resistance. They have free-cutting properties for use on lathes and mills.

3 series (e.g. 3103 – manganese is the principal alloy) are rarely used, as they are generally not heat treatable but are slightly stronger than pure aluminium.

4 series (silicon is the principal alloy) are typically used in welding wire, brazing alloys where their lower melting temperature than other alloys is useful.

5 series (e.g. 5251, 5083 – magnesium is the principal alloy) are of medium strength, have good welding properties and are corrosion-resistant. They work-harden easily. Some grades are used for architectural purposes and have good properties for anodizing.

6 series (e.g. 6063, 6082 – magnesium and silicon are combined in this alloy) are of medium strength, very good for intricate components and are corrosion-resistant. Some alloys in this series have good electrical conductivity.

7 series (e.g. 7178 – zinc is the principal alloy) are some of the highest strength aluminium alloys and are used for highly stressed parts.

Steel

En 1 steel has a good machinability and produces finely finished components that will be used for light duty. It should not be used for tubular and hollow parts such as pressurized containers. Case hardening is not recommended and should not be used when subject to severe impact. Machining index of 200 per cent.

En 3 is a general purpose mild steel that will take some cold deformation. It can be machined to close tolerance and has reasonable toughness and machining qualities. Typically used for shafts, hubs and good for welding. Machining index of 100 per cent.

En 5 is suitable for moderately stressed components. It possesses good toughness rather than high tensile strength and is used for nuts, bolts, brackets and levers. Machining index of 70 to 80 per cent.

En 8 steel is used instead of mild steel for improved properties and is good for general engineering parts such as motor shafts, crankshafts, con-rods, flanges, brackets and housings that are not subject to high stress or wear. Machining index of 68 to 72 per cent.

En 9 is used for springs, cylinders, crankshafts, keys, small gears and ball race housings where toughness is not required. It has good wear resistance. Normalized, it has a machining index of 50 to 60 per cent.

En 15 is used for shafts, crankshafts, con-rods, gas cylinders and flywheels. Normalized machining index of approximately 65 per cent.

En 16 has good high tensile strength and good shock resistance and is useful for low temperature components. Ideal for crankshafts, con-rods and high tensile nuts and bolts. Machining index of 40 to 50 per cent.

En 19 is good for high tensile strength applications and has good ductility, shock resistance, good wear resistance when hardened and tempered and can be used for nitriding. Good for high and low temperature use. Used for con-rods, gears, high tensile bolts, propshaft joints. Machining index of 40 to 50 per cent.

Stainless Steel
303Se has good machining properties, is free-cutting and does not seize easily. Machining index of 70 to 80 per cent.

304 does not corrode easily, has good mechanical properties and is widely available. Machining index of 40 to 50 per cent.

310S has very high resistance to oxidization and corrosion even at high temperatures. Good strength and mechanical properties and suited to fabrication. Machining index of 40 to 50 per cent.

316 has resistance to pitting and acid with high strength at high temperatures. Machining index of 40 to 50 per cent.

A-286 has high strength and corrosion resistance to 700°C. It requires heat treatment for machining. Used in gas turbines.

The applications given on the materials listed have been taken from various catalogues and from information supplied by a number of users. The machining ability constant is a notoriously unreliable guide because it depends on the type of machining operation, the speed and force of the cutting, how rigid it is and the type and shape of the tools. Each material's machining ability will be affected by its specific composition, its annealed treatment and the previous stresses a component has experienced. The values quoted for thermal expansion of materials are generally taken at 20°C unless otherwise stated. Different rates of thermal expansion will be obtained the closer a material gets to its melting point. Again, thermal conductivity will be affected by the composition of the material. Yield, proof and tensile strength figures will differ for different compositions and substantially higher or lower operating temperatures and will be affected by the quenching, annealing, normalizing and other treatments that have taken place, along with any work hardening that has been applied to the material. These strengths will also be highly sensitive to poor machining and some materials will prove to be notch sensitive. The tables shown in this book do not offer a definitive list for ductility, though this should be taken into account when specifying or choosing the most appropriate material. Ensure you rely on the manufacturers' or suppliers' information.

APPENDIX II – *see* page 233

APPENDIX III – COLOUR TEMPERATURE GUIDE FOR HEATING STEEL

Blood red = 560°C
Dark cherry red = 580°C
Medium cherry red = 675°C
Cherry red = 745°C
Bright red = 845°C
Salmon = 900°C
Orange = 940°C
Lemon = 995°C
Light yellow = 1,080°C
White = 1,200°C
Dazzling white = 1,290°C

APPENDIX IV – COLOURS FORMED DURING THE TEMPERING OF STEEL

Pale yellow = 220°C
Straw = 230°C – hammers, knives, lathe tools, scribers
Golden yellow = 245°C – taps, dies, reamers, drills
Brown = 255°C – punches
Brown with purple = 265°C
Purple = 275°C – chisels, screwdrivers, axes
Dark blue = 290°C
Bright blue = 300°C – rules, saws, springs
Pale blue = 320°C

Appendices V and VI – *see* pages 234 and 235

APPENDIX VII - FORMULAE

Volume of a cylinder = $(\pi \times$ Dia. sq. $\div 4) \times$ height
Volume of a circular cone = $(\pi \times$ Dia. sq. $\times$ height) $\div 12$
Volume of a sphere = $(\pi \times$ Dia. cu.$) \div 6$
Volume of a torus = $(\pi$ sq. $\div 4) \times$ centre Dia. of ring $\times$ height of the doughnut
Area of a triangle = (base $\times$ height) $\div 2$
Area of a circle = $\pi \times$ radius sq.
Area of a segment of a circle = radius sq. $\div 2 \times (((\pi \times \varphi) \div 180$ deg.$) - \sin \phi)$
Area of a hexagon = $(\sqrt 3 \div 2) \times$ across flats sq.

Celsius to Fahrenheit = $\times 1.8 + 32$
Fahrenheit to Celsius = $-32 \div 1.8$
Celcius to Kelvin = -273.15

Stress = load/area

Strain = total amount of stretch/total original length

The formula for thermal convection is: q = h $\times$ A $\times \Delta$ T
q = energy transfer rate
h = film coefficient
A = exposed surface area
Δ T = temperature difference of fluid surface

The formula for thermal radiation is: A = q $\div \varepsilon \times$ F $\times \sigma \times (T^4 - Ts^4)$
A = required area of thermal radiator
q = energy rate input to radiator
ε = emissivity (at maximum temperature)
F = view factor (perfect view = 1.0)
σ = Stefan-Boltzman constant $(5.669 \times 10^{-8}$ W/m^2 .K)
T^4 = temperature^4 (Kelvin)
Ts^4 = environmental temperature^4 (Kelvin)

CONVERSION FACTORS

1°C = 1°(K)Kelvin
1mm = 0.03937008in
1in = 25.4mm
1gram/cm cu. = 0.03612729 lb/in cu.
1lb/in cu. = 27.6799gram/cm cu.
1mm sq. = 0.001550003in sq.
1inch sq. = 645.16mm sq.
1cm cu. = 0.06102374in cu.
1in cu. = 16.38706cm cu.
1mega Pascal = 1 Newton/mm sq.
1lb/inch sq. (PSI) = 0.006894757 Newton/mm sq.
1Newton/mm sq. = 145.0377lb/in sq. (PSI)
1bar = 14.50377PSI
1PSI = 0.06894757bar
1horsepower = 745.6999 watts
1watt = 0.001341022 horsepower
1BTU/hour = 0.2928104 watts
1watt = 3.415179BTU/hour
1ounce = 28.34952grams
1gram = 0.03527396ounce
1mph = 1.609344km/h
1km/h = 0.6213712mph
1m/s = 2.236936mph
1mph = 0.44704m/s
1kilocalorie = 1.163watt-hour
1watt-hour = 0.8598452kilocalorie
1cal/cm−°C = 0.2watt/m−K
1ton/in sq. = 20MPa (N/mm sq.)
1cal/cm−°C = 0.2watt/m−K
1lb/in cu. = 0.286g/cm^3
1lb/ft cu. = 6g/cm^3
1lb/sq in. = 160Mpa $\times 10^{-3}$
1°ΔC = 1°Δ(K)Kolm

APPENDIX II – MATERIAL PROPERTIES

Material	Watts per m K conductivity at 20 deg.	x10-6/K of linear expansion	Deg. C Melting point	g/cm3 Density	MPa or N/mm2 UTS	MPa or N/mm2 Yield/proof stress
Argon	0.018	n/a	-189.3	0.0016	n/a	n/a
Air	0.026	n/a	-220	0.001293	n/a	n/a
Aluminium Alloy 2014 - T6	154	23	507	2.8	599-661	510-599
Aluminium Alloy 5251 - H3	112	24.1	568	2.64	275-330	179
Aluminium Alloy 6063 -T5	209	23.4	616	2.7	206	151
Aluminium Alloy 7075 - T6	130	22.1	532	2.8	530	440
Beryllium copper (25) CDA175	83.73	17	871	8.24	1006	854*
Brass Cartridge (70/30 brass) CDA260	120	19.9	915	8.53	303-896	76-448
Free Brass Cutting CDA360 (Cold worked)	115	20.5	885	8.49	385	310
Brass (red) CDA230	159	18.7	990	8.75	405-605	395
Brass (Yellow) CDA268	116	20.3	905	8.47	315	95
Naval Brass (Uninhibited) CDA464	116	21.2	885	8.41	483	400
Bronze (Aluminium) CDA613	60.62	16.2	1040	7.89	483-586	207-400
Bronze (Commercial) CDA220 H04	189	18.4	1020	8.8	420-510	370
Bronze (Penny) CDA405	165	17	1025	8.83	360	295-345
Bronze (Phosphor) CDA510	84	17.8	975	8.86	324-965	131-552
Carbon Fibre (SP Systems)	400	2.1	130†/3650	1.8	1500	270
Cast Iron (grey)	58	10.5	1200	7.25	179	65-293
Concrete	1	11	2577	1.8-2.2	12-50	14-70**
Copper	384	17	1068	8.93	300	200
Delrin - DuPont	0.36	84.6	175(82†)	1.41	75.8	89.6
Diamond (Natural)	2000	1.18	4027	3.51	1200-3000	2900
Freon	0.01	n/a	-36	0.00551	n/a	n/a
Glass (96% Silicate)	1.38	9	700	2.4-5	35-90	n/a
Gold (Annealed)	310	14.2	1063	19.29	'120' 220	205
Graphite	24-168	7.8	3800	1.77	2450	1270**
Helium	0.15	n/a	-272	0.000164	n/a	n/a
Hydrogen	0.18	n/a	-259	0.031	n/a	n/a
Kevlar - DuPont 49 Aramid	0.04	-2	149 †	1.44	3620	n/a
Leather(dry)	0.14	n/a	n/a	0.86-1.0	33-48	n/a
Magnesium LA 141 Ultralight	80	21.7	n/a	1.35	145	124
Magnesium M13210 HM21A T8	135	26.8	605	1.78	235	170
Magnesium M13312 HM31A T5	104	26	482-605	1.8	300	270
Mercury	8.3-8.7	61	-39	13.5	n/a	n/a
Polystyrene (Expanded) Standard	0.03	0.017	80-105	0.017- 0.03	35-60	17
Rubber (Butyl)	0.14	77	120	1.15	17-35	n/a
Rulon LR - (Maroon) widely used good anti-wear	0.33	112	-240>288†	2.25	13.8	n/a
Rulon F - (Green) abrasion resistance	0.37	112	-240>288†	1.89	8.3	n/a
Rulon J - (Gold) SS or Al/Aloy	0.29	117	-240>288†	1.95	13.8	n/a
Rulon XL - (Tan) Aluminium	0.24	115	-240>288†	1.97	11.7	n/a
Rulon 641 - (White) Stainless Steel 316/304	0.37	88	-240>288†	2.25	19.8	n/a
Rulon 142 - (blue/green) Mild/Hardened Steel	0.69	88	-240>288†	3.16	21.4	n/a
Rulon 1410 (gold)	0.26	71	-240>288†	2.2	14.8	n/a
Silicon	83	4.2	1420	2.33	n/a	n/a
Silver	419	19.6	961	10.5	125	54
Silver CP grade (Cold Worked)	418.6	19.6	962	10.5	290	n/a
Sodium	126	71	97.8	0.97	n/a	n/a
Stainless Steel - 316 T3	16.3	16.2	1350	7.9	999-1275	723 -1034
Stainless Steel - 303Se T2	16.2	17.1	1350	7.9	861-1034	586-861
Stainless Steel - A-286 T3	12	17.5	1350	7.9	965-1172	792-999
Stainless Steel - 310s T3	14.2	15.5	1400	8	999-1275	758-1103
Steel En1 (30 carbon) 220M07 G12130	51.9	11.5	1480	7.82	540	417
Steel En3 (20 carbon) 070M20 (Mild Steel) G10200	49.8	12.4-13.4	1480	7.86	470	395
Steel En5 (30 carbon) 080M30 G10300	50.24	12.1-14.7	1450	7.83	485.39	303.3
Steel En8 (40 carbon) 080M40 G10400	51.92	11.3	1320	7.828	620	550
Steel En9 (55 carbon) 070M55	46.05	12.7	1430	7.83	639.83	330.94
Steel En15 (carbon manganese - High tensile) 150M36	46.05	12	1480	7.88	697.74	551.58
Steel En16 (manganese molybdenum) 606M36 G50460	48.14	12.3-15.0	1280	7.82	1920	1570
Steel En19A (1% chrome molybdenum) 708M40 G4140 ##	43.12	12.8-14.9	1430	7.845	**1657.49**	**1505.81**
Steel En19A (1% chrome molybdenum) 708M40 G4140 ##	43.12	12.8-14.9	1430	7.845	**694**	**562.61**
Steel En24 (1.5% nickel chrome molybdenum) 817M40 G10250	44.5	12.3	1280	7.85	745	470
Steel En30B (4.25% nickel chrome molybdenum) 835M30	30	12.6	1425	7.85	1378	1172
Steel En32 (carbon case hardening) 080M15 G10150	159.9	12	1490	7.87	420	350
Steel En44B 060A96 G10950	51.9	11	1290	7.87	720	590
Steel En34 (2% ni-ch-moly cch lower carbon)665M17	51	12	1290	7.87	655	380
Styrofoam IB	0.028	0.07	75	0.026	450kpa	200kpa
Titanium Ti-6AI - 4V (Annealed)	7.1	8.5	1700	4.43	940	860
Titanium Ti-5	7.8	9.4	1660	4.49	861	827
Water	0.6	0.00021g/mL	0	1	9(ice)	n/a
Wood (Oak) Quercus Robur	0.17	5[54]	n/a	0.91{0.39}	110 [3]	**43 [7.5]
Wood (Norway Spruce) (Picea Abies)	0.2	5.8[34]	230†	0.3	70	21
Wood (Balsa) (Ochroma pyramidale)	0.05	1.1 #	n/a	0.16	13.5 [0.8]	**12.7

# = Change due to moisture will be more apparent	† = Max sevice temperature	n/a = not available or applicable
## = different heat treatment (**BBB**)	†† = at 4 degrees celsius	volumetric expansion
* = 0.1% Proof Stress	[n] = across grain	'000' = soft
** = compressive yield strength	0.0 {0.0} = Late {Early} - wood	

APPENDIX V – METRIC, IMPERIAL, BA, STANDARD WIRE
GAUGE AND DRILL NUMBER COMPARISON CHART

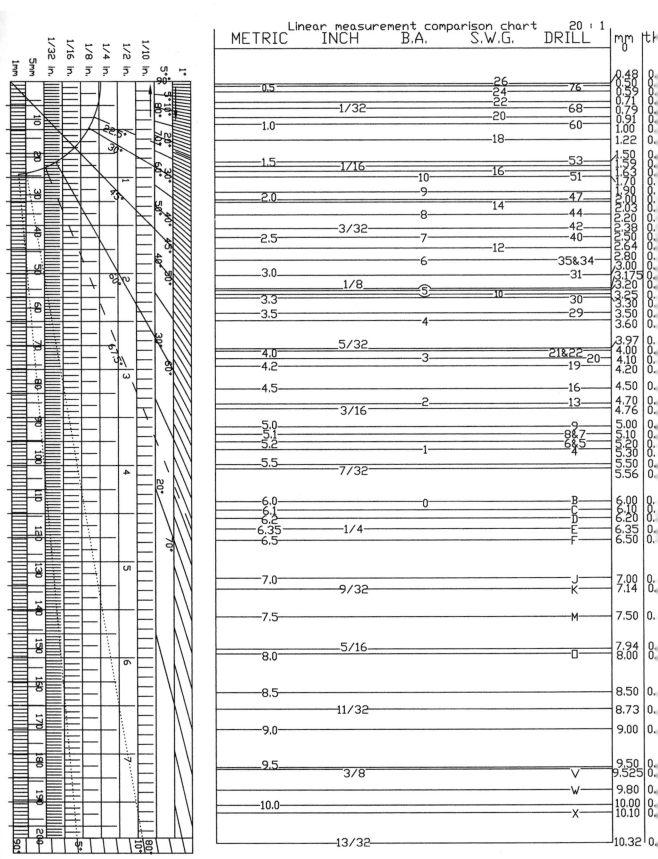

APPENDIX VI – BS 308 PRIMER
GLOSSARY AND DEFINITIONS

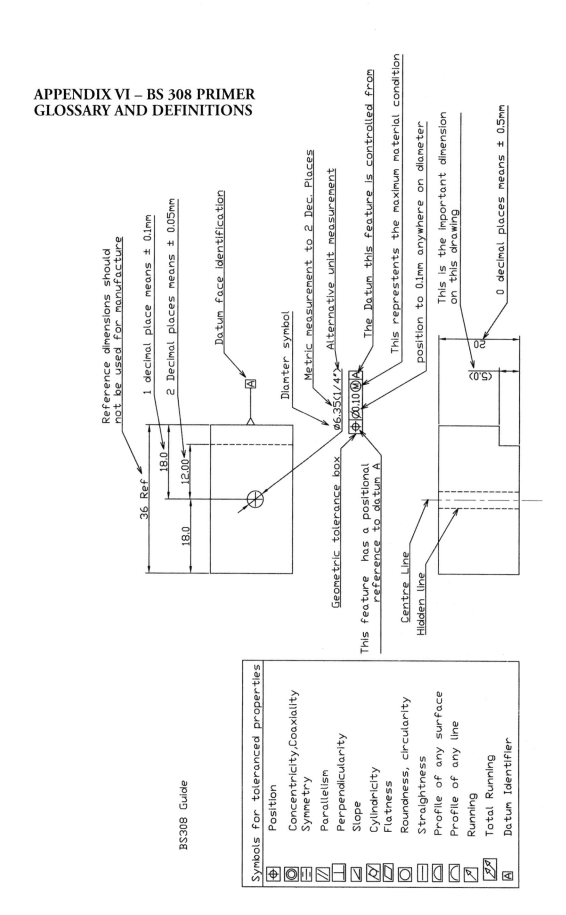